D0855225

BRUCE HOOD

The Domesticated Brain

A PELICAN INTRODUCTION

PELICAN BOOKS

Published by the Penguin Group

Penguin Books Ltd, 80 Strand, London WC2R ORL, England

Penguin Group (USA) Inc., 375 Hudson Street, New York, New York 10014, USA

Penguin Group (Canada), 90 Eglinton Avenue East, Suite 700, Toronto, Ontario, Canada M4P 2Y3 (a division of Pearson Penguin Canada Inc.)

Penguin Ireland, 25 St Stephen's Green, Dublin 2, Ireland (a division of Penguin Books Ltd)

Penguin Group (Australia), 707 Collins Street, Melbourne, Victoria 3008, Australia (a division of Pearson Australia Group Pty Ltd)

Penguin Books India Pvt Ltd, 11 Community Centre, Panchsheel Park, New Delhi – 110 017, India

Penguin Group (NZ), 67 Apollo Drive, Rosedale, Auckland 0632, New Zealand (a division of Pearson New Zealand Ltd)

Penguin Books (South Africa) (Pty) Ltd, Block D, Rosebank Office Park, 181 Jan Smuts Avenue, Parktown North, Gauteng 2193, South Africa

Penguin Books Ltd, Registered Offices: 80 Strand, London WC2R ORL, England

First published 2014
004

Book design by Matthew Young
Set in 10/14.664 pt Freight Text
Typeset by Jouve (UK), Milton Keynes
Printed in England by Clays Ltd, St Ives plc

ISBN: 978-0-141-97486-6

www.pelicanbooks.com
www.penguin.com

For my mother, Loyale Hood

Contents

PREFACE

The Incredible Shrinking Brain

Over the last 20,000 years, the human brain has shrunk by about the size of a tennis ball.[1] Palaeontologists found this out when they measured the fossilized skulls of our prehistoric ancestors and realized they were larger than the modern brain. This is a remarkable discovery by any standards, since for most of our evolution the human brain has been getting larger.[2] A shrinking brain seems at odds with the assumption that advancing science, education and technologies would lead to larger brains. Our cultural stereotypes of large egg-headed scientists or super-intelligent aliens with bulbous craniums fit with the idea that smart beings have big brains.

Small brains are generally not associated with intelligence in the animal kingdom; this is why being called 'bird-brained' is regarded as an insult (though in fact not all birds have small brains). Animals with large brains are more flexible and better at solving problems. As a species, humans have exceptionally large brains – about seven times larger than should be expected, given the average body size. The finding that the human brain has been getting smaller over our recent evolution runs counter to the generally held view that bigger brains equal more intelligence, and that we are smarter than our prehistoric ancestors. After all, the complexity of modern life suggests that we are becoming more clever to deal with it.

Nobody knows exactly why the human brain has been shrinking, but it does raise some provocative questions about the relationship between the brain, behaviour and intelligence. First, we make lots of unfounded assumptions about the progress of human intelligence. We assume our Stone Age ancestors must have been backward because the technologies they produced seem so primitive by modern standards. But what if raw human intelligence has not changed so much over the past 20,000 years? What if they were just as smart as modern man, only without the benefit of thousands of generations of accumulated knowledge? We should not assume that we are fundamentally more intelligent than an individual born 20,000 years ago. We may have more knowledge and understanding of the world around us, but much of it was garnered from the experiences of others that went before us rather than the fruits of our own effort.

Second, the link between brain size and intelligence is naïvely simplistic for many reasons. It is not the size that matters but how you use it. There are some individuals who are born with little brain tissue or others with only half a brain as a result of disease and surgery, but they can still think and perform within the normal range of intelligence because what brain tissue they do have left, they use so efficiently. Moreover, it's the internal wiring, not the size, that is critical. Brain volume based on fossil records does not tell you how the internal microstructures are organized or operating. Relying on size is as ridiculous as comparing the original computers of the 1950s that occupied whole rooms with today's miniature smartphones that fit into your pocket but have vastly more computing power.

Structural arguments aside, why would such a vital organ as the human brain, one that has been expanding for the majority of our evolution, suddenly begin to shrink around 20,000 years ago? One theory is related to nutrition. As we shifted away from being hunter-gathers living off meat and berries to subsistence farmers who cultivated crops, the change in our diet over this period might have been responsible for the brain change. However, this seems unlikely. Farming only recently arrived in Aboriginal Australia and yet they also experienced the same decline in brain size over this period. Also, agriculture first appeared in Asia around 11–12,000 years ago, well after the human brain had begun to change.

Environmental scientists point out that once the climate warmed up around 20,000 years ago, marking the beginning of the end of the Ice Age, we no longer needed large bodies to carry heavy loads of fat reserves. This might have led to a corresponding decline in brain size. Big brains require lots of energy, so a decrease in body size would have enabled our ancestors to downsize the brain, too. But that explanation fails to address similar periods of climate change that also took place during the 2 million years when hominid brains were still increasing in size.

One other theory about the reason for the shrinking brain may seem quite absurd – that the human brain is smaller because we have become domesticated. Normally, the word 'domestication' brings to mind washing machines, ironing, mortgages, weekend barbecues and family. While domestication has come to refer to all of these facets of modern home life, it was originally a biological term to describe artificial selection and breeding of plants and animals. Charles Darwin was fascinated by domestication and indeed many of his arguments for his theory on the origins of species were based on the effects of selective breeding by man of plants and animals, as an insight into the way that natural environments select to favour some individuals for reproduction over others. But unlike natural selection, domestication is not blind: with the invention of farming and animal rearing some 12,000 years ago, humans deliberately manipulated the selection process of both plants and animals to eventually modify the various species they wanted to exploit. We wanted animals to be more docile so that we could rear them more effectively. We bred aggression out of them by selecting individual animals that were easier to manage and, in doing so, we changed the nature of their behaviour.

This is how we also began to domesticate ourselves to live together in larger cooperatives. It was *self*-domestication because, unless you believe in divine intervention, humans have not been reared by a higher being who selected only some of us for reproduction. Rather, we have been self-regulating so that certain traits that were more acceptable to the group than others proliferated because individuals who possessed them were more successful in surviving and having children. In this sense we have been self-domesticating through the invention of culture and practices that ensure that we can live together.

Something about the domestication process produces profound and lasting physical changes. When wild animals are domesticated, their bodies and brains change along with their behaviour.[3] The brains of all the roughly thirty animals that have been domesticated by humans have decreased in volume by about 10–15 per cent in comparison with their wild progenitors – the same reduction observed over the last 1,000 generations of humans.

This effect on the brain has been observed in a set of experiments into selective breeding. In the 1950s, Russian geneticist Dmitri Belyaev began a programme of research to see if he could domesticate the Siberian Silver Fox.[4] Unlike modern dogs that are descendants of a strategy of selective breeding of wolves, most foxes have remained wild. Belyaev thought that domestication depended on temperament. Only those foxes that were less aggressive and less likely to run away when approached by the experimenter were chosen for breeding. These animals were tamer because, coded in their genes, they had slightly different brain chemistry regulating their behaviour. After only about a dozen or so generations of selective breeding, the offspring were markedly more docile. But they experienced significant physical changes too. They developed a white patch on their foreheads, were smaller than wild foxes and, like many dogs, had floppy ears. As Darwin noted in *On the Origin of Species*, 'not a single domestic animal can be named which has not in some country drooping ears'. They also had smaller brains.

Breeding for tameness as opposed to aggression means selecting for physiological changes in the systems that govern the body's hormones and neurochemicals. One possible mechanism to explain smaller brains is that individuals who are more passive may naturally have lower levels of the hormone testosterone. Testosterone is associated with aggression and dominance behaviour in animals, but its anabolic properties also play a rôle in body size by making muscles and organs larger and stronger. It also increases brain size. Sex-change individuals undergoing hormone treatment to facilitate the change to the opposite sex were found to have either increased or decreased brain volume depending on which hormone they were taking.[5]

Not only does domestication in animals lead to smaller brains, but it also changes the way they reason. Brian Hare, a leading expert on animal behaviour at Duke University, has shown that domesticated dogs in comparison to wild wolves are much better at reading the social signals of others. We humans can easily read the direction of another person's gaze to indicate where their focus of attention is. As we shall see in later chapters, it is a social skill that is present in young infants but becomes more sophisticated with the more social interactions we have as we develop. Domesticated dogs can also read human social signals[6] such as gaze and even the uniquely human gesture of pointing with the hand, whereas wolves and most other animals are generally baffled or indifferent.

Most fascinating is the change in dependence. Wolves will persist in trying to solve a difficult task through cunning, using different solutions, whereas the dog will typically give up earlier and try to recruit the help of its master. Domestication not only makes the animal more socially skilled but also more dependent on others. Over the years, several of the domesticated foxes from the breeding farms in Russia escaped back into the wild, only to return days later, unable to survive on their own.[7] They were dependent on those that had raised them.

Could domestication apply to human evolution as well? As a young researcher at Harvard, Hare went to a dinner where Richard Wrangham, a distinguished primatologist in the anthropology department, described how bonobos, the pygmy chimpanzee species famous for their sexual promiscuity when resolving disputes, were an evolutionary puzzle with a set of unusual traits not found in chimpanzees. Hare realized this was true of silver foxes as well. The more he looked at the similarities between domesticated animals and bonobos, and the way they differed from chimps, the more the evidence seemed to support a hypothesis that this subspecies of primate had become self-domesticated. The way their social groups had evolved placed a greater emphasis on social skills and conciliation rather than aggression. If this was true for bonobos, then why not humans?[8] After all, humans are also primates that have evolved the most remarkable capacity for social interaction. Hare would later write, 'human levels of flexibility in using others' social cues may have evolved in the human lineage only following the emergence of species-specific social emotions that provide motivation to attend to other individuals' behaviour and, subsequently, their communicative intent during purely cooperative interactions'.[9] In other words, the need to be more sociable by cooperating had altered the operations of the early hominid brain.

This is an old idea that has lately been re-seeded with new research and potential mechanisms. It first appeared under the guise of Social Darwinism in the nineteenth century – the idea that there were selective pressures that emerged from living together that changed the nature of the individuals. At first glance, it seems a bizarre hypothesis that living peacefully together caused the human brain to change, let alone shrink. After all, humans have been civilized for much longer than 20,000 years, with many earlier examples of societies, religions, art and culture. The recent discovery of stone artefacts on the Indonesian island of Flores that date to one million years ago tentatively indicate that an early hominin ancestor, *Homo erectus*, had inhabited the island.[10] If correct, that means that *Homo erectus* must have had considerable seafaring skills that would have required the cognitive capacity and social cooperation to coordinate such a voyage on early rafts because the land masses were separated by substantial amounts of open sea.[11]

Clearly our ancestors were cooperating and communicating well before the end of the last Ice Age. But there was a rise in the population around this time that could have increased pressure to adapt to cohabitation in larger groups.[12] Analysis of our species's history reveals that the world's population rose significantly in three continents well before the Neolithic period began around 12,000 years ago.[13] When the ice sheets covering the northern continents began to melt around 20,000 years ago, the demographics of our species changed rapidly, creating social environments that required increased levels of skills to navigate. The process of selection for social traits must have started when our hominid ancestors first began cooperating hundreds of thousands of years ago, when domestication first began to appear, but it could have undergone a sharp acceleration when they settled down to live together after the last great Ice Age.

Strength and aggression were advantageous for hunter-gatherer existence, but in these settled communities cunning, cooperation and trade were necessary. Humans now would have had to keep cool heads and even tempers. Those who prospered in this new selective environment would pass on the temperaments and social abilities that made them skilled at negotiation and diplomacy. Of course, there have been extreme violence and wars in the modern era and we have developed technologies to kill each other in vast numbers, but modern combat is typically orchestrated by groups; brute individual aggression was more prevalent in the smaller hunter-gatherer tribes of our prehistory.

By self-domesticating, we have been changing our species by promoting genes that produce relatively slowly developing brains in comparison to bodies. This would mean longer periods of development and social support that would have incurred more parental investment. It would require mechanisms that modulate temperaments and teach children how to behave in socially appropriate ways. Humans who lived together more peacefully in settled communities reproduced more successfully. They acquired skills that enabled them to cooperate, share information and eventually create our cultures.

Modern civilization arose not because we suddenly became more intelligent as a species, but rather because we learned to improve upon technologies and knowledge that we inherited by sharing information that was a by-product of domestication. Long childhoods were useful for transferring knowledge from one generation to the next, but they originally evolved so that we could learn to get along with everyone in the tribe. It was the drive to learn to live together in harmony that enabled collective intelligence to thrive, not the other way around. By sharing knowledge we became more educated, not necessarily more intelligent.

In 1860, two intrepid Victorian explorers named Robert Burke and William Wills set out on an expedition to cross the Australian continent from Melbourne in the south to the Gulf of Carpentaria in the north – a distance of 2,000 miles. They were successful in reaching the north coast, but on the return journey they both succumbed to starvation. Burke and Wills were educated modern men, but they did not know how to survive in the Outback. They were living on a plentiful supply of freshwater shellfish and a plant known as 'nardoo' that the local Aboriginals ate. However, both contain high levels of an enzyme that destroys vitamin B1, which is a vital amine (hence 'vitamin') essential for life. By ignoring the traditional Aboriginal method of roasting the shellfish and wet grinding and then baking the nardoo, which neutralizes the toxic enzyme, Burke and Wills had failed to capitalize on the ancient cultural Aboriginal knowledge. They did not die because of a lack of things to eat, but of beriberi malnutrition. Aborigines did not know about vitamin B1, beriberi or that intense heat destroys enzymes; they just learned from their parents the correct way to prepare these foods as children – no doubt knowledge that was acquired through the trial and error of deceased ancestors. Their cultural learning had provided them with critical knowledge that Burke and Wills lacked. As the two explorers' fates show, our intelligence and capacity for survival depends on what we learn from others.

Learning through domestication entails the transfer of knowledge and practices that are not always evident, either in terms of their purpose or origin. In the case of roasting Australian 'bush tucker', this practice was how to safely prepare food, but other examples include hunting and childbirth, both potentially life-threatening activities associated with folk wisdom. Of course, much folklore also contains superstition and irrational beliefs, but as we will discover in the following chapters, there is a strong imperative to copy what those around you say and do, especially when you are a child.

As a developmental psychologist, it is my view that childhood plays a major role in understanding the cultural evolution of our species. I usually tell my students at the University of Bristol the oft-cited finding that animals with the longest rearing periods tend to be the most intelligent and sociable. They also tend to be found in species that pair-bond for life rather than those that have multiple partners and produce many self-sufficient progeny. So it should be no surprise that of all the animals on this planet, humans spend proportionally the longest period of their lives dependent on others as children, and then as parents investing large amounts of their time and effort in raising their own offspring. This is how our species has evolved.

Of course, that profile of extended parental rearing is not unique to humans, but we are exceptional in that we use childhood to pass on vast amounts of accumulated knowledge. No other species creates and uses culture like we do. Our brains are evolved for it. As the leading developmental psychologist Michael Tomasello once quipped, 'Fish are born expecting water, humans are born expecting culture.' Other animals have the capacity to pass on learned behaviours such as how to crack nuts or use a twig for prodding a termite hill, but none have the same ability to transmit wisdom that increases with complexity from each generation to the next. Our ancient ancestors may have taught their children how to make a simple wheel, but now we can teach our children how to build a Ferrari.

The ability to transmit knowledge requires communication. Other animals can also communicate, but only limited and rigid information; humans, with our unique capacity for generating language, can tell limitless stories – even fantastical ones that are literally out of this world. We can also speak, write, read and use language to reflect on the past or think about the future. And it is not just the complexity and diversity of human language that is unique. Language had to build upon an understanding and desire to share knowledge with others who, like us in the first place, were willing to learn. It required understanding what others were thinking. Communication is part of our domestication – we had to learn to live peacefully and cooperatively with others for the collective good by sharing resources that include knowledge and stories. We do not just educate our children – we also socialize them so that they become useful members of society governed by all the rules and behaviours that hold it together.

Of course, this does not mean that our species is necessarily peaceful. There are always tensions and struggles in a world of limited resources and individuals will group together to defend their position against members from another tribe. However, for the conflicts that arise between groups and among individuals, modern societies govern with a greater level of control through our morality and laws than previously experienced in our history. To be an accepted member of society, each of us must learn these rules as part of our domestication.

We are such social animals that we are completely preoccupied with what others think about us. No wonder reputation is paramount when it comes to feeling good about ourselves. The social pressure to conform involves being valued by the group because, after all, most success is really defined by what others think. This preoccupation is all too evident in our modern celebrity culture, and especially with the rise of social networking, where normal individuals spend considerable amounts of time and effort in pursuit of recognition from others. Over 1.7 billion people on this planet use social networking on the Internet to share and seek validation from others. When Rachel Berry, a character in the hit musical series *Glee*, about a performing-arts school, said 'Nowadays being anonymous is worse than being poor', she was simply echoing our modern obsession with fame and our desire to be liked by many people – even if they are mostly anonymous or casual acquaintances.

We have always preferred others for what they can do for us. In the distant past, it may have been the individual attributes of strength to bring home the bacon and fight off competitors or our capacity to bear and raise many children that were selected for, but those attributes are no longer essential in the modern world. In today's society, it is as much strength of character, intellect and potential financial earning that most regard as desirable traits. Top of the list of qualities most of us would like to possess is high social status, which explains why many individuals who already are well off in every other domain of their life still seek the attention of others.

What others think about us is one of the most important motivations for why we do the things we do. Some of us may have moments of blissful solitude when we escape the rat race of modernity and pressure to conform, but most inevitably return to seek out the company and support of others. Deliberate ostracization can be the cruellest punishment to inflict on an individual, short of physical harm. Like domesticated foxes that escaped into the wild, we invariably need to return to the company of others.

Why is the group so important and why do we care about what others think? *The Domesticated Brain* shows that we behave the way we do because of how our brains evolved to be social. For humans, being social requires skills of perception and comprehension when it comes to recognizing and interpreting the activity of others but it also requires changing our own thoughts and behaviours to coordinate with theirs so that we can be accepted. This domestication as a species took place over the course of human evolution as self-selecting mechanisms shaped social behaviours and temperaments that were conducive to living in communities, but we continue to domesticate ourselves during the course of our own lives and especially during our most formative years as children.

Our brains evolved for living in large groups, cooperating, communicating and sharing a culture that we passed on to our children. This is why humans have such a long childhood: during this formative period, our brains can become acclimatized to our social environment. The need for social learning requires babies to pay special attention to those around them but also enough flexibility to encode cultural differences over the course of childhood. This enables each child to recognize and become a member of its own group. A child must learn to navigate not just the physical but the social world by understanding others' unseen goals and intentions. We have to become mind readers.

We need to develop and refine skills that make us capable of reading others in order to infer what they are thinking and most importantly, what they think about us. Where possible, evidence from comparative studies is considered to reveal the similarities and differences we share with our closest biological cousins, the non-human primates. And of course, we focus on human children. Developmental findings that reflect the interplay between brain mechanisms and emergence of social behaviour are the key to understanding the origins and operations of the mechanisms that keep us bound together.

That analysis could rely solely on the costs and benefits of social behaviours, but then we would miss the important point that people are emotional animals with feelings. It is not enough to read others and synchronize with them in some coordinated tango to achieve optimal goals. There is also an imperative to engage with others through positive and negative emotions that motivate us to be social in the first place. Taking that perspective casts a better light on understanding why humans seem to behave so irrationally because sometimes they care too much about what others think.

One of the more controversial issues that *The Domesticated Brain* addresses is the extent to which early environments can shape the individual and even pass on some acquired characteristics to their offspring. For most Darwinians committed to the theory of natural selection, whereby the environment alone operates to select genes that confer the best adaptations, this idea sounds heretical. Yet we examine the evidence that early social environments leave a lasting legacy for developing our temperaments through what are known as epigenetic processes – mechanisms that change the expression of our genes that can affect our own children.

Every child at some point has been told that they must 'behave' and when they do not, they are 'misbehaving'. What parents really intend when they scold their children for misbehaving is that they must learn how to control their thoughts and actions that conflict with the interests or expectations of others. Self-control is a feature of our developing frontal lobes of the brain and is central to our capacity to interact with others. Without self-control we would never be able to coordinate and negotiate by suppressing the urges and impulses that could interfere with social cooperation. This capacity for self-control is critical when it comes to being accepted and without it we are likely to be rejected – labelled anti-social because we fall foul of the moral and legal codes that hold our societies together.

That danger of rejection is the flipside of the benefit of living in a group and the devastating consequence of becoming an outsider. Ostracism and loneliness not only register as pain in our brain but also make us both psychologically and physically ill. Rejection can make individuals behave in destructive ways not only against themselves when they self-harm, but also against others. We may be more connected through social networking on the Internet, but this digital village also makes it much easier to become isolated.

Considering the vast size of the different territories covered in *The Domesticated Brain*, from human evolution, brain growth and child development to genetics, neuroscience and social psychology, any attempt to bridge these regions will be ambitious – yet it is a goal worth pursuing. When we recognize the importance of others in shaping who we become and how we behave, we can begin to understand what makes us human.

CHAPTER 1

Navigating the Social Landscape

'Why do you need a brain?' Initially, this seems like a silly question with an obvious answer. 'You need a brain to stay alive,' is a fairly common response and indeed this is true.[1] You would be dead without your brain. When someone is 'brain dead', they lack the vital signs of breathing and a heart beat – functions that are automatically controlled by structures deep at the core of the brain. However, keeping you alive is neither the sole function nor responsibility of the brain. There are many other organs you need to keep you alive. There are also many living things that do not have brains, such as simple organisms like bacteria, plants and fungi.

When you take a closer look at our planet and consider all its different life forms, it soon becomes apparent that the original reason why living things evolved brains was for movement. Life forms that do not move or those that are swept around by the ocean currents or carried in the wind or even transported on or inside the bodies of other animals do not need to have brains. In fact, some start off with brains that they later abandon.

The best example of this is the sea squirt that begins life as a tadpole-like creature, swimming around the ocean in search of a suitable rock upon which to attach. It has a rudimentary brain to coordinate movements and even a simple eye spot to 'see', but when it finally attaches to the rock, it no longer needs to search for a home and so loses its own brain.[2] Brains are expensive things to operate so if you no longer need one, why keep it?

Arguably, the main reason that the brain evolved was to navigate the world – to work out where you currently are, remember where you have been and decide where you are going next. The brain interprets the world as patterns of energy that stimulate the senses, generating signals that stream up into our brain where they are analysed and stored. With experience, these patterns become learned so that the brain knows how to respond more appropriately in preparation for future encounters. As you progress up the tree of life to animals with increasingly complex brains, you find that they have a much larger library of patterns they have stored. This provides greater flexibility, giving the animal more skills and knowledge to deal with potential problems rather than being stuck with a limited set of actions. Without the ability to act, organisms would be completely at the mercy of the environment. They would be easy pickings for any predator, unable to forage or capture their own food and vulnerable to the elements. Some creatures live their lives like this – the inevitable food for others – but many evolved a brain to lash out at the world or scamper away if the threat was too fierce.

The human brain, on the other hand, is not just for solving practical problems of finding food and avoiding danger; it is also a brain exquisitely engineered to interact with other brains. It evolved to enable humans to seek out others who are similar to form social relationships. Many of its specialized operations address the complexities of the social spheres we inhabit. We require a brain with finely honed skills to process different individuals who may be family, friends, workmates or the multitude of strangers we encounter in everyday situations.

In our ancestral past, these encounters would have been few and far between, but in the modern era we need to be expert socializers. We need to recognize who people are, what they are thinking, what they want and how to cooperate – or not – with them. We have to read others in order to understand them. These social skills that may seem trivially easy for many of us turn out to be some of the most complicated calculations our brains can perform. Some people never master them, such as individuals with autism, and others lose these capabilities through the effects of damage and disease to their brains. Our brain may have initially evolved to cope with a potentially threatening world of predators, limited food and adverse weather, but we now rely on it to navigate an equally unpredictable social landscape. The human brain enables each of us to learn about, and from, each other – to become domesticated.

Our brain is equipped with the mental machinery to live together, to breed, to raise our children and to pass on information about how to become a valued member of society. Many animals live together in groups but only humans have brains that enable them to transmit knowledge and understanding from one generation to the next in a way that is unparalleled in the animal kingdom. We can learn the rules about how to behave in ways that are acceptable to the group. We can adopt a moral code about what is right and wrong. We raise our children not only to survive to an age where they are capable of reproduction themselves but also to benefit from the collective wisdom of others that is passed on as culture.

Some scientists are not so impressed with our human capacity for culture. Primatologist Frans de Waal argues that other animals also have culture because they can learn from others and transmit that learning on to the next generation.[3] Famous examples of animal culture include the nut-cracking chimpanzees of Africa[4] or the Japanese macaques who wash the sand off sweet potatoes given to them by researchers.[5] In each case, juveniles have learned to copy what they observed in older animals. Just recently, three different neighbouring communities of chimpanzees living in the same habitat of the Ivory Coast have been shown to have distinct patterns of tool use to crack open Coula nuts.[6] At the beginning of the season, when the nuts are hard, stone hammers are used by all; but later in the year when the nuts become softer and more amenable, one group switches to using wooden hammers or tree anvils. A third group makes this transition more rapidly. These distinct behaviours can only be explained by learning, as all tools are potentially available to each group.

There can be little quibble with the evidence in these examples of animal tool use, but this imitation is not the same as the cultural transmission that occurs when we teach our children. There is no solid evidence that cultural learning in animals has led to technologies that are improved upon, modified and developed from each new generation to the next. We return to this issue in later chapters when we explore how human children not only copy an adult's tool use to solve a problem, but also faithfully copy rituals that have no objective purpose; something that animals have not been observed doing.

The debate about culture in animals is contentious, and our concern here is instead with what animal studies teach us about how humans are different. By addressing social mechanisms that most of us take for granted because they seem so natural and effortless, we examine how our brain has evolved to become domesticated, concentrating on childhood because this is when the major building blocks of domestication are laid down. But first, we must consider some of the basic processes that shaped the human brain to be capable of learning to become social.

Evolution in a nutshell

The only reasonable answer to where our brain came from is evolution by natural selection as famously described by Charles Darwin in the nineteenth century. Following from Darwin, most scientists today believe that life started out billions of years ago as simple chemical compounds in a primordial soup that somehow (we still don't really know how) developed the ability to copy themselves. These early replicators were the precursors of life, eventually developing structures called cells. Clusters of these cells in time collected together, evolving into the ancient life forms known as bacteria that are still with us today.

Everywhere you look, from the deepest oceans to the highest mountains, from the frozen tundra to the desert furnace, or even in the volcanic acid pools that would strip the skin off most animals, you will find bacteria that have adapted to the most extreme conditions that can be found on our planet. Through the process of evolution, life forms continued to change and develop in ways that enabled them to survive different environments. But why evolve?

The answer is that there is no reason behind evolution, it just happens. Organisms evolve as adaptations to aspects of the environment that pose threats to survival and, more importantly, reproduction. When living organisms reproduce, their offspring carry copies of their genes. Genes are chemical molecules of deoxyribonucleic acid (DNA) encoded within each living cell that carry information about how to build bodies. The biologist Richard Dawkins famously likened bodies to simple vehicles for carrying genes around.[7] Over time, various mutations arise spontaneously in the genes, creating slightly different bodies that lead to variations in the repertoire of adaptive fit. Some of these variations produce offspring who are better suited to the changing demands of the environment. The offspring who survive go on to produce further offspring with those inherited characteristics which worked so well, and so that adaptation becomes programmed into the genetic code that is passed on to future generations.

Through the relentless culling of those least suited for survival as natural selection dictates, the tree of life sprouted ever-increasing branches of diverging species that gradually evolved adaptations better suited to reproduce. This continuous winnowing process produced the diversity and accumulation of complex life forms that now fill the various niches of our planet – no matter how unforgiving they may be.

The ability to move our bodies purposefully around the world may have been the initial reason that brains evolved, but clearly humans are more complex than sea squirts.

Complexity suggests purpose and goals whereas evolution is a blind process driven by an automatic selection that chooses the best variations that spontaneously arise as part of the copying process. It is for this reason that Dawkins calls evolution 'The Blind Watchmaker'.[8] Any complexity that an animal has is usually sufficient to deal with the problems they need to solve. However, as environments are constantly changing, animals need to keep evolving or become extinct – which, when you look back on the course of life on earth, has happened to most. One estimate[9] suggests that of all the species that have lived on the Earth since life first appeared here some 3 billion years ago, only about one in a thousand is still living today – that's only 0.1 per cent.

There may be some controversies over the exact details and dates of this brief history of evolution, but as far as science is concerned, the origin of the species by natural selection is the only game in town when it comes to explaining the diversity and complexity of life on Earth. Whether we like it or not, we are related to all other life forms – including those with and without brains. However, human brains have enabled us, like no other animal on the planet, to bend the rules of natural selection because of our capacity to change our environment. That manipulation is largely a product of our domestication as a species.

The cost of big brains

When you consider that humans can survive in the hostile environment of outer space, where there is lethal radiation and no atmosphere, it is clear that we have considerable capacity for adaptation. When our early hominid ancestors first appeared some 4–5 million years ago, the environment was undergoing rapid changes and fluctuations that required a brain capable of versatility to deal with complex situations.[10] We have brains that can think up solutions to overcome the physical limits of our bodies so that we can live under water, fly through the sky, enter outer space and even bounce around on the surface of an alien planet that has no atmosphere suitable for life. However, the processing power to solve complex problems is costly.

The modern adult human brain weighs only 1⁄50 of the total body weight but uses up to 1⁄5 of the total energy needs. The brain's running costs are about eight to ten times as high, per unit mass, as those of the body's muscles; and around 3⁄4 of that energy is expended on the specialized brain cells that communicate in vast networks to generate our thoughts and behaviours, the neurons that we describe in greater detail in the next chapter.[11] An individual neuron sending a signal in the brain uses as much energy as a leg muscle cell running a marathon.[12] Of course, we use more energy overall when we are running, but we are not always on the move, whereas our brains never switch off. Even though the brain is metabolically greedy, it still outclasses any desktop computer both in terms of the calculations it can perform and the efficiency at which it does this. We may have built computers that can beat our top Grand Master chess players, but we are still far away from designing one that is capable of recognizing and picking up one of the chess pieces as easily as a typical three-year-old child can. Some of the skills we take for granted depend on deceptively complex calculations and mechanisms that currently baffle our engineers.

Each animal species on the planet has evolved an energy-efficient brain suited to deal with the demands of the particular niche in the environment that the animal inhabits. We humans developed a particularly large brain relative to our body size but we don't have the largest brain on the planet. Elephants can make that claim. Nor do we have the largest brain to body ratio. The elephant nose fish (which looks like an aquatic elephant) has a much larger brain to body size ratio than the human. Despite the recent brain shrinkage described earlier, the human brain is still around five to seven times larger than expected for a mammal of our body size.[13] Why do humans have such big brains? After all, big brains are not just metabolically expensive to run but they pose a considerable health risk to mothers. You only have to look in a Victorian graveyard to see the number of mothers who died during childbirth as a result of haemorrhaging and infection to understand why giving birth can be such a dangerous event.[14] Babies with large brains have large heads, which makes them more difficult to deliver. This became a particular problem during the evolution of our species when we started to navigate the physical world on two legs. When we began to walk upright with our heads held high, this increased the danger of childbirth but, inadvertently, this risk may have been responsible for a significant change in the way we looked after each other. It could have contributed to the beginning of our domestic life as a species.

Although most mammals are up and running about pretty soon after birth, human babies require constant care and attention from adults for at least the first couple of years. The newborn brain also has to undergo considerable growth. At birth, it is nearly twice as large as that of a chimpanzee when you take into consideration the size of the mother, but still only about 25–30 per cent the size of the adult human brain; a difference that is mostly made up within the first year.[15] Both our large growing brains and immaturity have led some anthropologists to claim that humans are born too early.[16] It has been estimated that instead of the standard nine months, humans would required a longer gestation period of eighteen to twenty-one months to be born at the same stage of brain and behavioural maturity equivalent to a chimpanzee newborn.[17] Why do humans leave the womb so early?

We do not have records of the brains of our ancient ancestors because the soft tissue deteriorates in the ground whereas bony skulls fossilize, and we can use these to estimate how big the brain they housed must have been. One of our first ancestors in the hominid tree of evolution appeared on the planet around 4 million years ago. *Australopithecus* or *southern ape* was very different from all the other ape species because it was able to walk upright on two legs. We know this because of the bone structures of their fossilized skeletons and the analysis of footprints that were preserved in the mud. The most famous fossil of *australopithecus* is called

'Lucy' after the Beatles song 'Lucy in the Sky with Diamonds' that was playing on the radio when she was unearthed in Ethiopia in 1974. Although Lucy was a young woman when she died, she was only about the height of a modern three- to four-year-old child and had a brain the size of a human newborn. She had long arms and curved fingers, so she was probably making the transition from living in trees to living on the land. One reason that Lucy may have come down from the trees was that the climate in Africa changed so that there was less jungle and more grassland savannahs. On a savannah, you are more vulnerable to attack from predators and so moving across flat land is much easier and faster on two legs than scrambling around on all fours like other apes.

Most of us take walking for granted, but moving on two legs is remarkably difficult. Just speak to any engineer who has tried to build a walking robot. We are familiar with science-fiction robots walking on two legs, but the reality is that this is extremely complex and requires sophisticated programming as well as a very level surface. This is because two legs provide only two points of contact with the ground, which is very unstable. Just try getting two pencils to balance against each other and you get the idea. Even big feet don't make it much easier. Add to that the problem of co-ordinating the shift in weight to lift one foot off the ground and then transfer that weight to the other foot as you stride. No wonder walking is considered to be a form of controlled, continuous falling forwards.

Walking and running were both adaptations to the changing environment of the flat grasslands but they came at a cost. First, even a nimble early hominid was not going to be able to out-run sabre-toothed cats or bears, so they had to be able to out-smart animals that were physically much larger, stronger and faster. Hominids had to evolve a brain not only capable of bipedal locomotion but one that was strategic enough to avoid capture. Second, when our female ancestors began to stand upright, this changed the anatomy of their bodies. For efficient movement on two legs, the hips have to be within a certain size, otherwise we would end up waddling like a duck – which is not the ideal way to run to catch prey or avoid being eaten. So there was adaptive pressure to keep the hips from becoming too wide, which, in turn, meant that the pelvic cavity, which is the space in between the hips, could not become any larger. The pelvic cavity determines the size of the birth canal, which effectively determines the size of the baby's head that a mother can deliver.

Up until 2 million years ago, the relative brain size of our hominid ancestors was the same as that of the great apes today. However, something happened in our evolution to change the course of our brain development, which grew significantly larger. Human brain-size increased to be 3–4 times larger than the brain of our ancestral apes.[18] As our head started to increase in size to accommodate our expanding brains, this put pressure on hominid mothers to deliver their babies before their heads got too big. However, this is not a problem for our nearest non-human cousins, the chimpanzee. In terms of movement, chimps do not naturally walk upright and so did not develop a narrow pelvis. Their birth canals are large enough to give a relatively easier birth to their babies, which is why chimpanzees waddle when they do try to walk upright. They usually deliver by themselves in less than 30 minutes, whereas human delivery takes considerably longer and is most often assisted by other adults.

This problem of birthing big-brained babies in slim-hipped mothers is known as the 'obstetrical dilemma' and until recently was the accepted account of why human infants are born so early relative to other primates. However, anthropologist Holly Dunsworth at the University of Rhode Island has argued that another reason why our infants are born so early is that mothers would starve if the gestation period was any longer.[19] Pregnancy is incredibly demanding on the mother in terms of the energy required to support both herself and the rapidly growing foetus. In primates and across other mammals, there is a reliable relationship between the relative size of the newborn compared to the mother that indicates that each species's delivery date represents the point where the energy demands of the foetus begin to exceed what the mother can safely provide.[20] Bigger foetuses require more energy. Dunsworth argues that pelvic size is not the only problem, but rather feeding babies without starving the mother is why humans are born prematurely.

What is undeniable is that human childbirth is not easy. One of the more intriguing ideas about the evolution of humans and their growing brains is that the difficulty and dangers posed by childbirth could have led to the development of assisted deliveries and ultimately contributed to the evolution of human domestication.[21] Humans needed help in order to give birth, which means that the onset of midwifery may have contributed to the social development of our species. No other animal has assisted childbirths and this unique feature which appeared early in our history may have been significant in shifting our species towards greater prosocial interactions. Other primates give birth relatively quickly in trees or bushes by themselves. It is possible for humans to give birth alone, and many do, but it is not the norm and especially not for first-time mothers, who typically experience longer, more painful labours. Assisted childbirth is part of our domestication. Having other members of the group present would have helped to protect against predators and reduce the stress of the situation by offering reassurance as well as provide physical assistance in actually delivering the baby.

Assisted childbirth could have been an early behaviour that fostered the right conditions for compassion, altruism, trust and other social exchanges that would become the behavioural foundations of our cultural domestication. Even if helping a mother to deliver entailed nothing more than being present to obscure, distract or confuse a potential opportunistic predator, these behaviours could have been the basis for reciprocal relationships with others in the group. Moreover, the stress and relief associated with a potentially dangerous birth could have triggered emotions that foster motivations to shape behaviours. Those who sought and offered assistance could have passed on such traits to their own offspring, thus increasing the likelihood of this cooperative behaviour becoming an established social pattern in the species.

In the same way that domesticated dogs seek assistance, when faced with a problem, our earliest ancestors began to look to others for help. Childbirth as shared emotional experience in the evolution of social behaviour may be highly speculative, but for anyone who has witnessed a birth for the first time, the extent of the experience is unexpected, surprisingly emotional and often beyond reason and control, suggesting that it triggers behaviours that lie deep in the history of our species to help others.

Brain size and behaviour

Considering all the problems that giving birth to big brains seems to entail, we are still left with the question, 'Why did our ancestors evolve much larger brains about 2 million years ago?' One possibility that is consistent with the argument we began with is that a larger brain enabled animals to move around and keep track of where they have been.[22] If you look at the animal kingdom, different patterns of feeding are related to different brain sizes. Primates who eat mostly fruits and nuts have larger brains than those primates who eat only leaves. Leaves are readily available in predictable locations and so require less foraging. Primates who live mostly on leaves have to consume much larger volumes of these low nutritional foods that then have to be broken down by enzymes in the stomach. This is why leaf-eating primates have much larger guts for fermenting the material. It also explains why they have to spend most of their day sitting around and simply eating and digesting.

In contrast, fruits and nuts are more nutritious but they are also sparse, more seasonal and more unpredictable. Coming down from the trees and learning to walk upright meant that foraging over greater distances by our ancestors would become the typical behavioural pattern. Bigger brains would have been necessary to find higher value nutritional foods that would have been necessary to maintain a bigger brain.

This is why fruit-eating primates have to travel much further to satisfy their dietary needs. They also have much smaller guts and proportionally larger brains. Their habitats are more extensive and require greater navigational skills so they are generally more active. Take spider monkeys and howler monkeys, two closely related species that live in the tropical rainforests of South America. The diet of the spider monkey is 90 per cent fruit and nuts, whereas howler monkeys live mostly on the rainforest's canopy leaves. This difference in diet and the need to forage could explain why the spider monkey's brain is proportionally twice the size of the howler monkey's, with a corresponding greater level of problem-solving abilities.

But our early ancestors were not simply foraging for nuts and berries – they were beginning to process food and carcasses with rudimentary stone tools. Animals with large brains are better tool users and humans are experts who far exceed any of the tool-making skills of other animals. Even making the earliest simple stone tools required special skills that are uniquely human. The anatomy of the human hand and the brain mechanisms that coordinate dexterity enabled our ancestors to hold a flint in one hand and knap it into the right shape with the other – a skill so far not observed in non-human primates.[23] Animals also tend to fashion tools from what is immediately available and abandon them soon after, whereas our ancestors hung on to their manufactured tools, carrying them around for future use. That requires a level of knowledge, expertise and intelligent planning to develop technology unprecedented in the animal kingdom – one notable exception being the sea otter that is said to carry a stone in its pouch that it uses for cracking seashells!

As unique as human tool use is, a significant increase in brain size occurred between 2 and 1.5 million years ago, and yet the oldest stone tools are between 3 and 2 million years old, predating the expansion of the hominid brain.[24] There have been considerable developments in the sophistication of the tools following the expansion of the brain, but the invention of tool technology itself probably did not seem to depend on the significant increase in brain size.

Another class of explanation is required to explain the need to develop larger brains but one that can include changing patterns of both food exploration and hunting. Early humans not only foraged but they also increasingly hunted, which meant that they had to travel further and they had to collaborate. They had to understand each other and cooperate to satisfy mutual goals. They had to navigate a social environment as much as a physical one and this social environment would soon get crowded.

One big family tree

The fossil record shows that modern humans are the last survivors of a branch of the evolutionary tree or genus of the apes known as *Homo* that emerged during a period of time known as the Pleistocene that began some 2.5 million years ago. Recent discoveries in Kenya reveal that this was a crowded time, with multiple hominid species co-existing.[25]

Other members that would emerge later out of this branch include *Homo hablis, Homo erectus, Homo heidelbergensis, Homo neanderthalensis* and *Homo floresiensis*, nicknamed the 'hobbit' because of its small stature. All have become extinct, with *floresiensis* being the last to disappear, possibly as recently as 12–15,000 years ago. We are *Homo sapiens* ('wise man'), who first appeared in Africa some 200,000 years ago.[26]

In addition to evidence based on the fossil record, scientists have been able to reconstruct our human past by analysing the human DNA genome and looking for common sequences that reveal our relatedness. By using statistics, they can work out how long it took patterns to deviate to reconstruct our ancestry. One type of DNA, which is found outside the nucleus of cells known as mitochrondial DNA (mtDNA), has been particularly useful because it provides a way of tracing the history of our species and identifying the spread of humans across the globe. In a female, mtDNA is stored in her eggs and mutates at a different rate than cellular DNA. This difference in mutation rates enables researchers to establish various lineages back into the dark prehistory of our species. In 1987, researchers published results of mtDNA analysis and reported evidence that there was a common ancestor who must have lived in Africa around 200,000 years ago who was the ancestor for all modern humans.[27] As this was based on the female mtDNA that was passed on to the thousands of her grandchildren, this hypothetical mother became known as 'mitochondrial Eve'. Just recently, scientists have been able to extract DNA from *Homo neanderthalensis* to determine that we are related to this extinct subspecies, while also revealing a bit of a prehistoric scandal.

Homo sapiens and *Homo neanderthalensis* were known to be living close to each other in the same parts of Europe at around 40,000 years ago. Eventually, *Homo sapiens* became the last survivors. The more ancient *Homo neanderthalensis*, who first appeared on the scene 700,000 years ago, disappeared in Europe and it was assumed that they had been out-manoeuvred or wiped out by the *Homo sapiens* from Africa through competition for resources. However, it would now appear that there was some 'Pleistocene hanky panky' going on, as British-born palaeoanthropologist Ian Tattersall called it, referring to the genetic evidence of interbreeding.[28] Analysis published in 2011 revealed that, on average, billions of people outside Africa have about 2.5 per cent of Neanderthal DNA in their genome.[29] Of course, we cannot know whether this interbreeding was cooperative or forced, but it does paint a completely different picture of our species.

Homo psychologicus – the social brain hypothesis

Evolutionary psychologist Robin Dunbar at Oxford University has argued that humans evolved large brains to enable them to live in large social groups.[30] Domestication in recent human history may have triggered a reduction in brain size over the last 20,000 years, but brains had to initially grow larger during the much longer extent of hominid evolution over the past 2.5 million years in order to live in social groups. This idea, known as the *social brain hypothesis*, argues that communal living required the development of large brains to navigate the social landscape but not all animals that live in large groups have particularly big brains. If that were so, we would expect the wildebeest that migrate in vast numbers across the plains of Africa to be particularly cerebrally well endowed – which they are not. They form large herds but they are not organized and coordinated by complicated social relationships. So merely living as part of a social group does not adequately explain the increased size of brains. Rather, you have to look at the nature of the social interaction of animals that live in groups to understand why big brains confer social adaptation.

UCLA anthropologist Joan Silk has studied the social organizations of different apes and monkeys and thinks that it is the ability to recognize the relationships between other members, or 'third-party knowledge' – a sort-of 'he knows that she knows' type of understanding – that is the critical skill for living in social groups.[31] Many primates are sensitive to such third-party knowledge. Upon hearing the distress call of an infant monkey, wild vervet monkeys hidden in the bushes will turn towards the mother and the direction of the call, which shows they recognize the mother–infant relationship. In chimpanzees, males form dominance hierarchies that confer all the advantages of fathering more offspring. These chimp gangs are based on allegiances formed by pretenders to the throne who recruit followers through social interactions in much the same way individuals form gangs to rule the school playground. Once in place, the new top boss or 'alpha male' has the pick of the females, but he will tolerate attempts to mate from those who helped him establish the new regime.

If today's non-human primates engage their social skills for power struggles, then it is likely that early hominids did the same. To support his social brain hypothesis, Dunbar analysed the relative brain size of many different animals and discovered that those with proportionately the largest brains are the ones that live in larger structured groups and possess more social skills. Primates in these groups have a larger repertoire of calls that enable them to communicate more complex information, a feat that requires larger brains.[32]

This relationship between brain size and social behaviour is found throughout the animal kingdom. It is not only true for social animals such as elephants but also sea-dwelling mammals such as dolphins and whales. It is also true in the bird world. A good case in point is the *Corvidae* family of crows, jays and magpies. Caledonian crows have bigger brains than the larger chicken and not surprisingly they are also considerably smarter. In fact, when faced with puzzles that are suitable for birds, Caledonian crows outperform many primates, which is why they have been called *feathered apes.*[33]

Longer childhoods are another feature of social animals who invest time raising their young. A chicken is independent by four months after birth and reaches maturity by six months, whereas Caledonian crows are still fledgelings at two years and require continual feeding from the parents. This is why corvid parents pair-bond for life, because it is an evolutionary strategy for sharing the responsibility of raising offspring that take so long to mature. Bigger brains may provide these animals with more flexibility in their problem solving, but they need it to be able to provide for their demanding kids.

Cultural explosion

When our species appeared on the scene some 200,000 years ago in Africa, *Homo sapiens* lived in organized social groups, communicating through gesture and simple language to enable them to cooperate and coordinate. We know this because the ancestor to both *Homo sapiens* and *Homo neanderthalensis, Homo heidelbergensis,* who had been around for maybe as long as 1.3 million years, was already a skilled hunter. In Schöningen, Germany between 1994 and 1998, eight exquisitely fashioned wooden throwing spears measuring 2 metres long were found among the skeletons of twenty horses. They were carved so that the weight was towards the front of the spear, making it fly straighter, similar to the design of a modern javelin. As a boy scout, I unsuccessfully tried to make spears and I doubt many of us today would know what the optimum design is. The Schöningen spears date to around 400,000 years ago, proving that *Homo heidelbergensis* was sophisticated enough to make a weapon sufficiently lethal to bring down a larger animal. This technological advance could not have suddenly appeared but rather must have been passed on through social learning. Since horses are difficult to corner, a hunting party would be needed to coordinate the attack, suggesting they had the ability to communicate. Given their expert skills in hunting horses, *Homo heidelbergensis* proves that culture was already present before the appearance of *Homo sapiens* 200,000 years ago.[34]

Soon after the appearance of *Homo sapiens*, other examples of social learning and culture began to show up in the fossil record. Samples of haematite, a red iron oxide that can be used as pigment for body adornment, have been found in Zambian sites dating to around 160,000 years ago. Ceremonial burials including a man clutching the jawbone of a wild boar have been dated to around 115,000 years. Other graves of the same period contained beads. Why go to this effort unless there was some symbolic meaning for the objects?

As they rapidly spread geographically across the planet, *Homo sapiens* must have been equipped with a brain capable of much more culture than ever seen before. Based on a statistical analysis of the global data set of mtDNA sequences, it is believed that there was an increase in the *Homo sapiens* population around 100,000 years ago that would have produced a demographic that was ripe for enabling culture to flourish through the exchange of ideas and migrations of individuals.[35]

From around 100,000 to 45,000 years ago, there had been sporadic examples of cultural practices such as ceremonial burials and symbolic behaviour like art and body decoration. However, in Europe around 45,000 years ago, *Homo sapiens* became anatomically modern humans engaging in all the trappings of primitive civilization. They were as close to us today as we can find in terms of their bodies. They also behaved much more like us than any other ancestor. Around this time there was a cultural explosion as evidenced by the advances in tool technology, elaborate jewellery, symbolic sculptures, cave paintings, musical instruments, talismans and the spread of religious ceremonies and burials.[36] Each of these activities was undertaken for a purpose that required a level of social interaction far in excess of anything seen before or remotely present in the animal kingdom. Humans had clearly begun to trade, as many of the raw materials for the artefacts had been transported great distances. In other words, we were already becoming vain. Art and jewellery are primarily made to be seen and admired by others. Making jewellery and creating art took considerable time and effort and would only have been undertaken and appreciated for the social value such activities conveyed. Burials and religious ceremony reflect an awareness of death and thoughts about the afterlife and creators. It may be true that some primates show the behavioural signs of mourning their dead, but modern humans are the only species that engage in death rituals.

The psychologist Nick Humphrey has suggested that it would be more appropriate to call our species *Homo psychologicus* (psychological man), given the ability of *Homo sapiens* to read minds – not in any supernatural psychic way, but simply by imagining what someone else is thinking and predicting what they may do next.[37] You need to be able to read others if you are a member of a species that has evolved to co-exist and, more importantly, cooperate. You also need these skills if you are producing helpless infants who need childcare and shared rearing. In order to make sure that you have enough resources for yourself and any offspring, you must be able to understand and anticipate the intentions and goals of other members of the group.

This is particularly true of primates who engage in deception and coalition formation, sometimes called 'Machiavellian intelligence' after the Italian Renaissance scholar who wrote about how to govern through cunning and strategy.[38] This ability requires a set of social skills known as 'theory of mind' in the psychological literature and represents a powerful component of social intelligence.[39] When you have a theory of mind, you are able to mentally put yourself in another's shoes to see things from their perspective. This enables you to keep track of others, to second-guess their intentions, to outwit them and to exchange ideas. As we will read in later chapters on child development, theory of mind has a protracted progress and for some unfortunate individuals remains impaired, which presents a considerable hurdle in communicating with others.

The chattering brain

One uniquely human social skill that we regularly use for problem solving is language. Although we sometimes talk to ourselves, the primary purpose of language is to communicate with others. We learn to speak by listening to others, and if we were raised in an environment where we heard no language, then all the evidence indicates that we could not learn to speak normally at a later age, no matter how much training and effort we put in. There is something in our biology that dictates that we must be exposed to language at a critically early age to acquire it.[40] Even learning a second language becomes increasingly harder as we age, indicating that there is a biological window of opportunity for language acquisition.

Just about every facet of human activity involves language, whether it is work, rest or play. No other animal on the planet communicates like we do. They may have squawks, barks, grunts, squeals, snorts, screams, cries, hoots and all manner of noises, but the information they are communicating is extremely limited and rigid. Despite what Walt Disney and other animators would like us to believe, animal communications are nothing more than elaborate signalling systems to convey one of four simple messages:

> 'Watch out, there's trouble about.'
> 'Back off, man, I mean business.'
> 'Come and get it, there's food over here.'

Or more often than not,

> 'Come and get it, ladies, I'm over here.'

Animal communication is primarily for the four Fs of fleeing, fighting, feeding and fornicating – basic drives that keep us alive long enough to pass on our genes by reproduction. Humans also spend a considerable amount of time communicating on these very topics but when we communicate, there is nothing we better like to do than talk about others. An analysis of typical conversations in a shopping mall revealed that two thirds of the content was related to some social activity – who's doing what with whom.[41] Human communication is not restricted to biological drives that are necessary for survival and reproduction. We can talk about the weather, politics, religion and even science. We can pass on opinions, instructions and all manner of other high-level, complex information, though in all likelihood our initial communications when language first appeared were probably directed to the same four Fs that were necessary for survival. After all, human communication is complicated and difficult to execute and therefore must have evolved for a good purpose.[42]

Why can't we talk with the animals? First, we are the only primates with the motor machinery that enables us to vocalize the controlled sounds that form the building blocks of speech.[43] Most notably, unlike other primates, we have a descended larynx. The larynx or 'voice box' serves a number of roles. As we exhale, the air passes by the vocal cords that vibrate to create sound in the same way that blowing across a blade of grass produces a quacking sound. Changing the shape of the mouth, tongue and lips as well as controlling our breathing can further modify these sound segments to produce the differing vocalizations. The other main role of the larynx is to close up in order to protect us from inhaling food, but it does not begin to descend in the human until around three months of age, which explains why babies can swallow and breathe at the same time when they are breastfeeding.

With our descended larynx, we have a much longer vocal tract, enabling us to produce a much greater variety of sounds. Coupled with this extra-long sound pipe, we also have greater muscular control over our lips and tongues compared to other primates, which is why human speech is physically impossible for other animals. But that physical limitation is not the only reason that animals do not speak.

They simply don't have the right brains for it. Karl Lashley, the American psychologist, originally proposed in 1951 that the unique basis of human speech must involve brain circuitry responsible for sequencing movements.[44] In recent years, this hypothesis has gained support from the discovery of the FOXP2 gene that governs the embryonic development of brain structures that support speech production. Even if animals could control the required movements, linguist Noam Chomsky emphasizes decoding the underlying structure of language itself as requiring specialized brain mechanisms that humans alone have evolved.[45] The major difference between our language and the social communication of other animals is that we have a system of grammar – words and rules that can combine together to generate an unlimited number of new sentences about anything. Most of us are not even aware that we are using these rules. As native speakers, we can spot that there is something wrong with the utterance 'complex human is language' because it does not follow the rules, but very few of us know exactly what these rules are. Before we discovered the rules of language, humans were speaking grammatically.

Language is also a symbolic system, which means we use sounds to stand for something. In speech these are the words, but before there were words there must have been specific sounds that we learned to associate with meaning. Animals can also learn to associate sounds to stand for things if they are trained to do so. They can even learn to associate gestures with meanings. There are some famous cases of chimpanzees that learned sign language, but this is not something they can do spontaneously. It requires lots of training with rewards and they cannot make up new sentences as easily as children do.

There is something special about human language in both production and understanding that other animals just do not get because it was never part of their evolution. Our capacity for language is arguably the major species-specific ability that catapulted modern humans into an unparalleled league of social interaction. It has not always been like this. A hunter-gatherer ancestor did not wake up one day and blurt out to the rest of the tribe, 'Let's go hunting.' Our language must have evolved into the complex behaviour that is universally enjoyed today. Some argue that evolution cannot explain something as complex as language but it is precisely because of that complexity that language had to evolve gradually by natural selection. In the same way that the eye is a complex biological adaptation that could not have suddenly appeared from a one-off mutation, the same must be true for language.

Babies do not need to be taught how to speak; most children are fluent by three years of age, irrespective of where they grow up in the world, so long as there are people around speaking to them. The grammars of industrialized societies are no more complex than those of so-called primitive tribes and all languages share the same underlying linguistic rules that were only relatively recently discovered. Language can also be knocked out by certain head injuries, it activates specific networks of neural circuitry in the brain and some language disorders are genetically transmitted. Taken together, these facts indicate that the development of language belongs more to the realm of human biology than cultural invention, which is why language has been called an instinct.[46] Language not only enabled humans to pass on information, but it allowed us to domesticate our children by instructing, scolding and encouraging those ideas and behaviours that would be most suited to getting on with others peacefully.

The architecture of the mind

Many scientists believe that language did not suddenly appear but rather must have evolved from a number of different sub-skills – almost like making a new machine by recycling other parts. Evolutionary psychologists Leda Cosmides and John Tooby propose that much of the mind must also be considered like a toolbox that has accumulated specialized skills over the millennia to deal with specific problems.[47] Like every other aspect of the human body, they argue that the brain must have evolved to solve problems through the process of gradual adaptation. As Cosmides and Tooby quip, 'the human brain did not fall out of the sky', ready prepared to address all of life's problems. Rather, it must have evolved in stages, dealing with one set of problems at a time. As humans evolved increasingly more complex lives, we also had to evolve new behaviours that provided the best opportunity for reproduction. We needed to find the best mate, refine attentive social skills and learn what was necessary to be accepted.

With these sorts of recurring problems, humans evolved a repertoire of coping skills that are passed on in our genes. Our ability to navigate, count, communicate, reason about the physical properties of objects and interpret expressions are just some of the candidate functions that might be part of our evolved behaviours. These can all be found in humans across the planet today, irrespective of where they live. If these functions are universal and largely independent of the culture or society, then this strongly suggests that they are wired into our biology and transmitted by our genes. However, this is where the theoretical arguments take place. To what extent is a particular human attribute an evolved adaptation and to what extent has it been created and transmitted in recent evolutionary history by culture? Is jealousy a cultural artefact of prevailing sexual attitudes or could it be something that conferred an adaptation in our evolutionary past? Even though we cannot go back to see how our ancient ancestors evolved, we can look for clues that support the idea that functions we possess are the legacy of natural selection.

Human evolution took place over millions of years and must have been gradual for a number of reasons. First, as an organism evolves from simple to more complex activities, the types of problems it encounters over time will change, spurring on the necessity for further adaptations. The complexity of the brain could not have resulted from one massive mutation in our DNA. Rather, the complexity would have had to emerge as each successive version of ancestors had to deal with a new set of problems. Second, adaptation works for solving specific problems, so a brain that was not especially equipped to deal with a problem would not be selected for. In effect, the brain had to have a collection of specialized problem-solving solutions rather than being a good all-rounder. If the brain had only been a good all-rounder problem-solving system, then it could never be as efficient as one made up of multiple specific skills. Different problems require different solutions with tailored mechanisms. In other words, a jack-of-all-trades is a master of none.

One way to imagine the mind is as a Swiss Army knife with lots of blades that perform different functions. You have blades for removing stones from a horse's hoof (who uses that these days?), corkscrews, scissors and an assortment of other bespoke blades. In the same way, the brain has specific functions such as language, spatial navigation, face processing, counting and so on. If our mind was like our metaphorical knife but only had one general-purpose blade good for cutting but not good for opening bottles, then we would be limited in dealing with specific problems. For example, vervet monkeys have evolved an alarm-call system to identify three different types of predator: snakes, eagles and leopards. Each predator requires a different course of action: either standing on hind legs looking down at the grass around them (snake), looking up in the air and diving into a bush (eagles) or climbing up a tree (leopards). Get the response wrong and the vervet monkey becomes dinner. This is why they instinctively respond to different alarm calls. A general-purpose 'Look out!' would not have been a good adaptation.

This evolutionary approach has led to the view that the architecture of the mind is not a general problem solver but rather a collection of systems dedicated to addressing specific problems. In the same way that dedicated mechanisms for solving recurrent problems during human evolution could have emerged through the process of natural selection, culture-gene approaches to understanding human evolution propose that our species possess mechanisms that reliably seek out cultural input.[48] In other words, there are genetic dispositions to learn efficiently. The reason for this is that culture changes faster than genes. Unlike examples of cultural learning in animals, humans continually refine, develop and expand on knowledge that is passed on. This is possible because we have brains that are evolved to learn from others. Our efficiency is guided not only from our capacity to communicate, but also by our biases to attend to specific aspects of others that signal who are most valuable as teachers. As we will learn in the coming chapters, babies are tuned into their mothers from the very start in a reciprocal relationship. But they also pay more attention to others who are older, who are the same sex, who are friendly and who speak the same language. Babies are born with dispositions encoded in their genes to learn from those who are going to be most useful to them in terms of acceptance by the group.

Cognition, cooperation and culture

Psychologist Mike Tomasello at the Max Planck Institute for Evolutionary Anthropology in Leipzig is one of the world's leading experts on what makes us human. He studies the development of children and how they compare to other primates. He believes that the traits that distinguish humans from our primate cousins are our capacity to think about others, cooperate with them and share ideas and behaviours. All of these are necessary for cultures to thrive. Human culture differs from any other social groups in the animal kingdom because there is a cumulative build-up of knowledge and technologies that is passed on from one generation to the next. With every generation, our world becomes more complex because we educate and share information by cooperation. In this way, knowledge and understanding 'ratchet up', with each successive generation expanding and improving the complexity and collective knowledge of the group.[49]

Other animals also live in groups and exhibit a host of social skills for working out what others are thinking, but these abilities are mostly restricted to situations where there is a potential fight or conflict. Most non-human primates are opportunists, only on the lookout for situations where they can take advantage of other members for either food or sex or to establish a better position in the dominance hierarchy. There are examples where chimpanzees will help others, but these are mostly situations where there is the potential for some personal gain.[50] In contrast, people will sacrifice personal gain for others. They will even spontaneously help strangers who they will never meet again. The capacity for altruism seems to be characteristically human. Examples of animal altruism are rare and restricted to those species that exhibit strong codependence, such as marmosets. In these cases, it is strategically in their interests to be promiscuously prosocial to increase their likelihood of breeding.[51]

Humans may be opportunists too, but all societies are held together by tacit assumptions of reciprocity and moral codes to prevent individuals taking advantage. These are the rules we abide by. Some of these codes are enshrined as laws. We enter into social contracts where we submit to authority or the state on the assumption that those who abide by the rules will benefit, whereas those who violate or break them will be punished. Members who benefit from these social arrangements do not even have to be family. Indeed, when you think about it, much human sharing of resources is altruistic – doing good deeds for the benefit of others who remain anonymous without necessarily benefiting ourselves.

No other animal on the planet behaves as altruistically as we humans do. Of course, there are some species, such as worker ants and bees, that make the ultimate sacrifice for the good of the nest or the hive when it comes under attack, but they do so because they are genetically closely related to those that benefit. Evolution has programmed their brains to be self-sacrificial. Humans are different. We cooperate with others because it makes us feel good. It is the thought of helping that is the reward, because we feel connected to the group. These feelings are the emotions that motivate us to be prosocial towards our fellow man (or woman) and fuel the drive towards altruistic collaboration, cooperation and ultimately human culture. However, we are not slavish drones that automatically bend over backwards to help anyone; we are always on the lookout for those who are trying to cheat the systems of reciprocity. We are inclined to lend a hand but we will seek retaliation if we believe we have been wronged. In order to make these sorts of decisions we have to have brains that are sophisticated enough to interpret others in terms of their motives, their goals and their affiliations.

What makes the human brain different?

For many animals, the problems of living long enough to reproduce were basic and immediate – how to navigate the world to find food, avoid harm and so on. Solitary animals figure these out for themselves because this is how they have evolved. Other animals that live in groups evolved the capability for coordination and cooperation for mutual benefit. For them, the environmental pressures they had to adapt to were not only physical, geographical or climate-based but also social. In a group, there would have been multiple potential mates competing to pass on their genes. This led to the evolution of social behaviours that increased the likelihood of successful breeding within a group.

This increase in social skills is considered one of the reasons that primate brains grew larger and why our species in particular have become the most skilled at interacting and learning from others. But then the human brain began to shrink again with the birth of large civilizations, when we started to live together more peacefully. It could be that humans went further than all other social animals by developing culture – the ability to communicate, to share ideas and knowledge, to engage in ritualistic symbolic activity and develop rules about how to behave for the benefit of the group. We had to learn to live together in greater harmony as our numbers started to increase. We needed to learn to become diplomatic. While physical environments tend to be static, social environments by comparison are constantly changing and providing considerable feedback, which in turn changes the dynamic of the interaction. In short, expertise in social interactions required considerable processing power and flexibility.

To enable humans to do this, we developed long childhoods to provide sufficient time and resources to ensure that our offspring were educated in the skills necessary for harmonious social living. Why else would humans have evolved into the species that spends the longest proportion of their lives dependent on adults? This amount of time was an evolutionarily big commitment for both parents and their offspring. With domestication came wisdom passed down the generations. We may have taught our own children some basics, but there was more to learn from the group. Our ability to communicate meant that our children could learn more about the world they needed to negotiate by listening to others without having to rediscover everything from first principles. But to benefit from that, the most critical knowledge they learned during childhood was how to be liked and valued by others – in other words, how to behave.

CHAPTER 2

Make Up Your Mind

According to available records, the youngest child convicted and executed for a crime in England was John Dean, aged around eight years. He was hanged in Abingdon for setting fire to two barns in the nearby town of Windsor in 1629. At the time, the age of criminal responsibility was seven years, at which point children were considered to be little adults. Indeed, this is how they were often portrayed in paintings from that period.

In Van Dyck's (1637) portrait of the children of Charles I, they look like miniature adults. Charles II, the boy in the painting, is only seven years old, yet he is shown with the posture of an adult, with feet crossed in a casual lean against the wall. The portrait reflects the prevailing attitude of the time that children simply lacked the wisdom of experience and that with training they would become acceptable to society. Like empty vessels, they needed to be filled up with information and instructed how to behave.

John Locke (1632–1704), the English philosopher, captured this view of the child as a blank canvas:

> Let us then suppose the mind to be, as we say, white paper void of all characters without any ideas. How comes it to

Figure 1: Children depicted as 'Little Adults'

> be furnished? Whence comes it by the vast store which the busy and boundless fancy of man has painted on it with an almost endless variety? Whence has it all the materials of reason and knowledge? To this I answer, in one word, from EXPERIENCE.[1]

Locke described the infant's mind as a '*tabula rasa*' or blank slate. Not only was the infant's mind considered empty, it was one that was faced with the daunting task of making sense of a complex and confusing new world of sensations and experience that the American psychologist William James would later describe in 1890 as a 'blooming, buzzing confusion'.[2]

However, Locke's blank slate view is not plausible, nor is the newborn's world completely confusing as James imagined. As the Prussian philosopher Immanuel Kant[3] (1724–1804) pointed out, blank slates would not work unless they were already set up to detect the structures of the world. There has to be some organization built in in order to determine what constitutes a pattern in the first place. Consider how complicated vision would be without some pre-existing knowledge. You cannot begin to understand the world around you unless you have some inkling of what you are looking for. In order to perceive the world, you need to distinguish objects from backgrounds and determine where one object begins and another ends. We rarely consider these as problems because vision is so effortless. It is only when you try to build a machine that can see that the difficulty becomes all too obvious.

In 1966, Marvin Minsky, one of the pioneers of artificial intelligence, is said to have asked one of his undergraduate students at MIT to 'spend the summer linking a camera to a computer and getting the computer to describe what it saw'. Presumably Minsky thought that the problem was easy enough that it should take the duration of a summer vacation for a student to solve. That was almost fifty years ago and thousands of professional scientists are still working on how to make machines see like humans.

Back in the 1960s, artificial intelligence was a new field of science that promised us a labour-saving future where robots clean the house, wash the dishes and basically perform all the mundane chores humans hate to do. Since then we have witnessed remarkable developments in computing and technology and there are certainly very smart vacuum cleaners and dishwashers. But we still have not been able to build a robot that perceives the world like a human. They may look human but they are unable to solve some of the simplest problems we take for granted, ones that most babies master before their first birthday.

Another reason why the blank slate cannot be true is because it turns out to be physiologically wrong. Our senses are pre-configured in anticipation of the sorts of signals that we can expect as babies. We do not have to learn to distinguish different colours, or that a boundary between brightness and darkness corresponds to an edge. If you measure from brain cells that react to sensation in unborn animals before they have had any experience of the external world, they will already respond to features that they have not yet encountered. Human newborns will show immediate preferences for

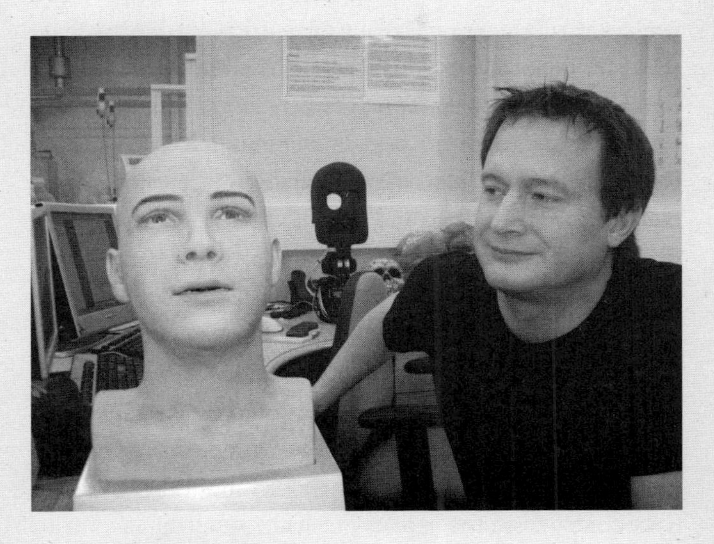

Figure 2: Spot the difference

some patterns before they have had any time to learn them, so their world is not totally confusing. These early capabilities show that there is considerable formatting in the newborn's brain that enables them to make sense of experiences.

Like a computer that you buy from a store, the brain already comes with an operating system installed. What you eventually store on it really comes down to what you do with it. Biology and experience work together to generate a developing mind, adapted to the external world. That process is one of discovery as each child goes about decoding the complexity of the world around them by using the tools that evolution has bestowed upon them.

Getting wired up

The brain of any animal is as complex as it needs to be to solve the world problems that the creature has evolved to accomplish. In other words, the more versatile an animal's behaviour, the more sophisticated their brain.[4] This versatility comes from the capacity to learn – storing memories as patterns of electrical connectivity in the specialized brain cells that alter in response to experiences. In the human adult, the brain is comprised of an estimated 170 billion cells, of which 86 billion are neurons.[5] The neuron is the basic building block of the brain's communication processes that support thoughts and actions.

Each neuron looks like a many-tentacled creature from outer space with a body from which branch thousands of receptors or *dendrites* that receive incoming signals from other neurons. When the sum of incoming nerve impulses reaches a critical threshold, the receiving neuron then discharges its own impulse down the axon to set off another chain reaction of communication. In this way, each neuron acts like a miniature microprocessor. The patterns of nerve impulses that spread across the vast network of trillions of neural connections are the language of the brain as information is received, processed, transmitted and stored in these networks. The presentations of experiences become re-presented, or *representations* – neural patterns that reflect experiences and the internal computing processes our brains perform when interpreting information.

One of the more surprising discoveries about brain development is that human infants are born with almost the full complement of neurons that they will have as an adult. Yet the newborn brain weighs about a third of the adult brain. Within a year, it is about three-quarters the size of the adult brain.[6] Connections form at a rate of 40,000 per second in the newborn, which is over 3 billion per day.[7] Eventually, the length connecting fibres will extend to an estimated 150–180,000 km – enough wiring within an individual human brain to circumnavigate the world's equator four times.[8] In fact, the bulk of the brain is mostly connections, with the neurons squeezed into 3–4mm of the outer layer that covers the surface of the brain, called the *cortex* after the Latin word for 'bark'.

These changes in connectivity enable the world to shape the brain by experience because experience keeps the neurons active through repeated mutual activation. This shaping process is called *plasticity* after the Greek word *plassein*, meaning 'to mould'. Synapses between cells that are in constant communication change in their sensitivity so that messages transfer more easily between them. At its most basic level, this is how information is stored in the brain – as changing patterns of neuronal activity. This crucial role for reciprocal neuronal activity is captured in the first of the neuroscientist's principles of plasticity, 'cells that fire together, wire together'.[9]

Most brain plasticity occurs during child development, with some areas continuing to change well into the late teenage years. The front part of the brain, associated with decision-making, does not become fully mature until the child reaches adulthood. Of course, there is plasticity in the adult brain as we constantly learn throughout our lifetime. However, connectivity in some brain systems seems to be time sensitive, requiring input much earlier in development. Remember neural activity is metabolically expensive. If neural connections are not active, then why keep them? In many ways it is similar to pruning your favourite rose bush. You cut away the weaker branches in order to allow the stronger branches to flourish.

These windows of opportunity, which are sometimes called *critical periods*, reflect the way that Nature has produced a brain that anticipates specific experiences at certain times; if these are denied or impoverished, there may be long-term impairment. This is true for the sensory systems such as vision and hearing, but as we will read in the next chapter, there appear to be critical periods for social skills as well. This loss of function due to deprivation is a second principle of plasticity, where you have to 'use it or lose it' when it comes to keeping neural mechanisms functional.

Core knowledge

In the same way that our brain is pre-configured to experiences before we have even had the chance to encounter the relevant sensations, some scientists believe that we are also wired to interpret the world in particular ways before we have had the chance to think about it. The speed at which babies acquire and understand aspects of the world around them before they are capable of comprehending spoken language indicates that they must be working things out for themselves. As adults, we take it for granted that the world is made up of objects, spaces, dimensions, plants, animals and all manner of complex ideas that we rarely take the time to consider because we have had a lifetime of exposure to them. But how do young babies come to appreciate these concepts in the absence of language? When a baby looks around its new blurry world, what does it make of it all? Even if they are learning by themselves, how do they know what to pay attention to and what is relevant? These sorts of problems have led to the proposal that some key components of understanding the world, especially those related to the physical nature of objects, numbers and space, must be programmed into the brains of infants from birth. But how do we know what babies are thinking when they are not even capable of telling us what is on their minds? The answer comes down to showing them magic tricks.

The reason that we find magic tricks so entertaining is that they violate our expectations. When a magician makes an object vanish into thin air, we are first surprised and then set about trying to figure out how they achieved the illusion. As adults, we know that a physical law has only apparently been violated because if we did not have that understanding, then we would not be surprised. That is why it is a trick. The same is true for babies. When they are shown magic sequences where objects appear to vanish, infants look longer. They do not applaud or gasp as an adult audience would, but they notice that something is not quite right.

This magic-trick technique, known as *violation of expectancy*, has spawned hundreds of experiments used to tap into the minds of infants who cannot tell us what they are thinking. Harvard psychologist Elizabeth Spelke has been using violation of expectancy to probe the rules infants apply when understanding the physical world.[10] From very early on, infants recognize that solid objects do not pass through other solid objects, move from one position to another without appearing in between, move by themselves unless contacted and nor do they dissolve or fall apart when touched. When we say that something is 'solid as a rock', it is so because it obeys Spelke's rules for physical objects. These rules do not have to be learned and for most objects that the child will encounter throughout the rest of their life, these principles will hold true, which is why they are referred to as *core knowledge*, because they are programmed into the mind from birth.

Of course, there are some exceptions to these rules, such as in the case of magnets, where an iron object will move in the absence of direct contact with another object. Soft bananas dipped in liquid nitrogen become hard as nails. These exceptions to the normal rules are enchanting because they violate our expectations of how physical objects should behave. Many toys that you find in science museums are counter-intuitive examples that amaze and amuse precisely because they do not behave like most ordinary objects.

It's alive!

Babies appreciate that people are also another type of object, but one with a special set of properties. For a start, people can move by themselves. If an inanimate object is left behind a screen then it should still be there unless someone has moved it. A person, on the other hand, can leave the room when you are not looking, so may not necessarily remain stationary when they are out of sight.[11] Also people do not have to move in a straight line. Five-month-olds who watched a video of a box sliding across a stage and passing behind two screens were surprised if it did not reappear in the gap in between. However, they were not surprised when a person moving across the same stage did not reappear in between the screens, suggesting that the infants could draw a distinction between how a box and a person can behave when moving in between screens.

Living things also move in particular ways. Objects that are not alive tend to move in a rigid way, whereas living things have 'biological motion' which is much more fluid and flexible. These types of movement are processed by neurons that are tuned to directions and speed in the visual area at the back of the brain known as *MT*. Biological motion is less rigid and activates a different region which is closer to the area behind your ears that is activated by faces. This area, the *fusiform gyrus*, also registers the shape of the human body, which suggests it might be a region that stores general information about others like us.[12] When we think of others, we expect them to be a certain shape and move in a certain way. By six months of age, babies are surprised to see a female who appears to have arms growing out of her hips that swing as she walks.[13]

How do babies decide what is human? We know that babies like to look at other people. They prefer biological motion at birth.[14] We also know they prefer the sound of the human voice and their mother's voice in particular.[15] They prefer the smell of their mother compared to the smell from another mother.[16] Just about every capability of the newborn's senses seems to be tuned into their mums.

Over time, infants gradually start paying attention to others and noticing what they are doing. When you think about it, the sheer volume of information contained in just a minute or two of a typical everyday action that an adult might perform is staggering.[17] Consider the individual steps involved in making a cheese sandwich. Every sequence requires complex motor skills that must be performed in a way that is beyond the capabilities of robots. Ingredients and implements must be retrieved from various locations in the kitchen and then prepared and assembled in the correct pre-planned order. There is no point trying to butter the slice of bread after the cheese has been inserted. How do babies begin to make sense of what they see when watching others? It turns out that in just the same way that baby brains are wired to chop language up into different segments, they are programmed to observe and learn different actions. Infants as young as six months are sensitive to the statistical regularities in action sequences and by ten to twelve months readily segment complex actions up into their constituent parts based on the flow of movements as they start and stop.[18]

So babies are the consummate *people persons* – they love to watch others. People are the most interesting objects to babies not only because they look and move in a particular way in complex action sequences but because they interact with them. Synchrony is critically important to establishing social interactions and babies are on the lookout for those who are tuned into them. As adults, we instinctively engage in these synchronized activities, often mimicking the baby in an attempt to capture their affections. Two-month-old infants will even treat non-living objects that act contingently as if they are alive and smile at them.[19] As they build up their models of what it is to be human, they are looking for evidence for those things that are most likely to be important for their survival and becoming increasingly more sophisticated in their decisions.

Thinking objects

Babies rely on faces, biological movement and contingent interaction to draw up a list of credentials that make something worth paying attention to. Any one of these attributes may signal that something is worth watching because they are starting to draw a distinction between the living and non-living world in terms of agency. Non-living things move because some force has acted upon them, whereas agents act independently for a purpose – they have goals. They have choices. When we understand that something has goals, we see it as intentional. We do this all the time with animals and our pets, when we give them human qualities using a cognitive bias called *anthropomorphism*, but we will even extend such 'humanness' to things that are clearly not alive, let alone possessed of minds.

Imagine three geometric shapes moving around a screen. A large triangle attacks a smaller triangle by banging into it and then corners a small circle inside a rectangular box. The circle moves around frantically inside the box as if trapped. The smaller triangle distracts the large triangle, allowing the circle to escape, and then closes the opening to the box, trapping the large triangle inside. The small triangle and circle rotate around each other in joy and then exit the screen. The large triangle proceeds to break up the box in a fit of rage. Hardly the script of a Hollywood blockbuster, but observers interpret this sequence as some sort of domestic dispute.[20]

This simple animation made by psychologists Fritz Heider and Marianne Simmel in 1944 demonstrates that humans anthropomorphize moving shapes that appear to be goal directed and generate rich interpretations consistent with social relationships. The philosopher Dan Dennett thinks that we adopt *an intentional stance* as a strategy to first look out for things that could be agents that could have consequences for us and then give them intentions.[21] When something has a face, moves as if alive or behaves in a purposeful way, we think that it has a mind that may have intentions directed towards us.

Attributing agency is something that babies also do from very early on. Based on the Heider and Simmel animation,

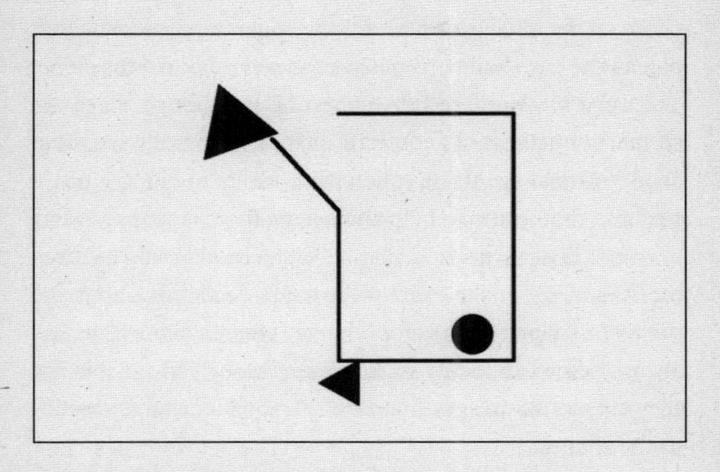

Figure 3: Scene from Heider and Simmel 1944 animation

infant psychologist Val Kuhlmeier showed infants a cartoon geometric red sphere, appearing to climb up a steep hill, that kept faltering and slipping down the slope.[22] At one point, a green pyramid shape comes along and pushes the sphere up the slope until it reaches the top. To most of us, this seems to be a case where the pyramid has helped the sphere up the slope. In a second scene, the red sphere is again trying to climb up the hill but this time along comes a yellow cube that blocks the path and then pushes the sphere down the slope. The cube has hindered the sphere. Even though these are simple animations of geometric shapes, we readily see them as intentional agents. A sphere that wants to climb a hill, a pyramid that wants to help and a cube that wants to hinder.

What is remarkable is that babies even as young as three months of age make exactly the same decisions about the different shapes.[23] They look longer when a shape that has always helped suddenly starts hindering. Already at this age they are attributing good and bad personality characteristics to the shapes.

Are you thinking what I am thinking?

We not only judge others by their deeds but we also try to imagine what is going on in their minds. How do we know what others are thinking? One way is to ask them, but sometimes you cannot use language. On a recent trip to Japan, where I do not speak the language, I discovered just how much I took communication for granted. But before language, there had to be a more primitive form of communication that enabled humans to begin to understand each other.

We had to know that we could share ideas, something that requires an awareness that others have minds and an understanding of what they might be thinking. The real quantum leap in the history of mankind that transformed our species was not initially language, but rather our ability to mind read.

Mind reading

I am going to surprise you with a little mind reading. Take a moment to look over the picture below, Georges de la Tour's famous painting 'The Cheat with the Ace of Diamonds' (1635), until you have worked out what is going on.

In all likelihood, your eyes were instinctively drawn to the lady card player at the centre of the picture and, from there, you probably followed her line of gaze to the waitress and then to the faces of the two other players. Eventually you will have spotted the deception. The player on the left is cheating, as we can see that he is taking an ace from behind his back to change his hand into an ace flush of diamonds. He waits for his moment when the other players are not paying attention to him.

How did I know where you would look? Did I read your mind? I did not need to. To fully understand de la Tour's painting, you have to read the faces and the eyes to work out what is going on in the minds of the players. Studies of the eye movements of adults looking at pictures of individuals in social settings reveal a very consistent and predictable path of scrutiny that speaks volumes about the nature of human interactions.[24] Humans seek out meaning in social

Figure 4:
'The Cheat with the Ace of Diamonds' by Georges de la Tour (1635)

settings by reading others, whereas another animal wandering through the Louvre Museum in Paris where de la Tour's masterpiece hangs would probably pay little attention to the painting let alone scrutinize the faces for meaning.

How do we begin to mind read? We start with the face. Initially we pay attention to the lady at the centre because the face is one of the most important patterns for humans. As adults, we tend to see faces everywhere – in the clouds, on the moon, on the front of VW Beetles. Any pattern with two dots for eyes that has the potential to look like a face is immediately seen as one. It may be a legacy of an adaptive strategy to look out for faces wherever they might be just in case there is a potential enemy hidden in the bushes, or it may simply be that because humans spend so much time looking at faces, we see them everywhere.[25]

When we look at faces, we concentrate on the eyes, which explains why this region is responsible for generating the most brain activity in observers.[26] Eyes serve several communicative purposes because they are directed to pick up visual information and, in doing so, reveal when and where someone's attention is focused. Gaze behaviour is also a precursor to communication, which is why we try to catch someone's eye before we strike up a conversation. By watching someone else's eyes, you can work out what is most interesting to them and when it is appropriate to speak. In face to face conversation, the person listening spends roughly twice the amount of time looking at the speaker, who will periodically glance at the listener, especially when they are making an important point or expecting a response.[27] We can gauge how much interest or boredom they are expressing and whether they have been registering the important messages by watching their gaze behaviour.

Not only do we seek out the gaze of others but it can also be difficult to ignore, especially if they are staring at us. This is why it is hard for soldiers standing to attention on a parade ground to maintain a fixed stare ahead of them when the drill sergeant, only inches away, stares at them and commands 'Don't eyeball me, soldier!' This focus requires a lot of discipline. Mischievous tourists notoriously try to make the guards on sentry duty outside Buckingham Palace lose their concentration by getting the soldier to look at them. Trying to avoid eye contact with someone in front of your face is almost impossible. Likewise, if a speaker you are listening to suddenly looks over your shoulder as if to spot something of interest, then you will automatically turn round to see what has captured their attention. This is because most of us instinctively follow another person's direction of gaze without even knowing it.[28]

Even infants follow eye gaze. When I was at Harvard, I conducted a study where we showed ten-week-old babies an image of a woman's face on a large computer monitor.[29] She was blinking her eyes open, staring either to the left or right. The babies instinctively looked in the same direction, even though there was often nothing to see.

Gaze monitoring works so well because each human eye is made up of a pupil that opens and constricts to allow varying levels of light into the eye, and a white sclera. This combination of the dark pupil set against the white sclera makes it very easy to work out where someone's eyes are pointing. Even at a distance, before we can identify who someone is, we can work out where they are looking. In a sea of faces, we are quickest to spot the face whose eyes are directed at us.[30]

Direct staring, especially if prolonged, triggers the emotional centres of the brain, including the amygdalae, which are associated with the four Fs.[31] If the other person is someone you like, the experience can be pleasing, but it is distressing if they are a stranger. Newborns prefer faces with a direct gaze[32] and, as we saw earlier, if you engage with a three-month-old baby by looking at them, they will smile back at you.[33] However, as children develop, patterns of gaze behaviour change because there are cultural differences in what is regarded as acceptable behaviour.

Cultural norms explain why staring at strangers is common in many Mediterranean countries, but makes foreign tourists often feel uncomfortable at being gawked at when on holiday.[34] Likewise, direct eye contact, especially between someone of lower status with someone of higher status such as a student and teacher in Japanese culture, is not considered polite. Japanese adults perceive direct gaze as angrier, unapproachable and more unpleasant.[35] Whereas in the West, we tend to think of someone who does not look you in the eye during a conversation as being shifty and deceitful.

When individuals from different cultures with different social norms get together, it can be an uncomfortable exchange as each tries to establish or avoid eye contact. This cultural variation shows that paying attention to another's gaze is a universal behaviour programmed into our brains at birth, but it is shaped by social norms over the course of our childhood. Our culture defines what are considered appropriate and inappropriate social interactions, influencing our behaviour through emotional regulation of what feels right when we communicate.

Mind games

By signalling the focus of another's interest, gaze monitoring enables humans to engage in joint attention. How many times have you been in the company of a bore at a party who is droning on and you want to leave with your partner or friend but cannot tell them directly? A roll of the eyes, raised eyebrows and a nod of the head towards the door are all effective non-verbal cues. Even if the other person is a stranger or does not speak your language, you would be able to understand each other without exchanging a word. Joint attention is the capacity to direct another's interest towards something notable. It is a reciprocal behaviour; you pay attention to what I am focused on and, in return, I pay attention to you. When two individuals are engaged in joint attention, they are monitoring each other in a cooperative act to attend to things of interest in the world.

Other animals, such as meerkats, can direct attention by turning their heads to signal a potential threat. Gorillas interpret direct gaze as a threat, which is why there is a sign at my local Bristol zoo where Jock the 34 stone 6 foot silverback male lives telling visitors not to stare at him. Jock pays attention to gaze as a source of threat, but we are the only species that has the capacity to read the meaning of the eyes over and beyond sex and violence (domesticated dogs being the notable exception that we described in the opening to the book). We use other people's gaze to interpret the nature of relationships. People who know each other exchange glances and those in love stare at each other.[36] This explains those awkward moments that we have all had when we exchange a glance or stare with a stranger in the street or, worse, in an elevator where it is difficult to walk away. Do I know you? Or do you want to be friendly or fight? At a party we can look around and work out who likes each other just by monitoring patterns of joint attention. This ability to work out 'who likes who' based on gaze alone develops as we become more socially adept. Six-year-olds can identify who are friends based on synchronized mutual gaze, but younger children find it difficult.[37] Joint attention in younger children and babies is really just from the child's own perspective. If it does not involve them, then they are not bothered. As children become socially more skilled at interacting with others, they start to read others for information that is useful for becoming part of the group.

Joint attention may have evolved as a means to signal important events out in the world in the same way that meerkats use it, but we have developed gaze monitoring into a uniquely human capacity to share interests that enable us to cooperate and work together. No other animal spends as much time engaged in mutual staring and joint attention as humans.

Gaze monitoring is also one of the basic building blocks for social cooperation. We are much more likely to conform to rules and norms if we believe that we are being watched by others. A poster with a pair of eyes reminiscent of George Orwell's 'Big Brother' makes people tidy up after themselves, follow garbage separation rules, voluntarily pay for beverages and give half as much again to charity boxes left in supermarkets.[38] Even though individuals may be totally alone, just the thought that they might be watched is enough to get most people to act on their best behaviour. Other people's gaze makes us become more self-conscious, prosocial and likely to conform.

It is notable that humans are the only primate out of the 200-plus species who have a sclera enlarged enough to make gaze monitoring so easy for us. In humans, the sclera is three times larger than that of any other non-human primate. If you think about it, evolution of the human sclera could not have been for the benefit of the individual alone. There would have been no selective advantage for me with my big white sclera unless there was someone else around to read my eyes. Rather, it had to be of mutual benefit to those who can read my eyes as well as myself.[39] It is only of use within a group that learns to watch each other for information.

When infants are learning words from an adult for things they have never encountered before, they listen out for the new name but also monitor where the adult is looking. In one study they were shown a new object and when they were looking at it the experimenter said 'Look at the toopa' but at the same time was herself looking into a bucket.[40] None of the children associated the word 'toopa' with the object they were holding. Children understand new words refer to new things but only those that are introduced in the context of shared joint attention.

By their first birthday, babies are constantly monitoring the faces of others, looking for information, and have even mastered the skill of pointing that can alert another to something of interest. Initially, babies point because they want something out of reach. Many primates raised in captivity do this as well, though it is more of an open-handed gesture to receive food. Apes also lack the hand anatomy that allows them to extend the index finger in the same way that humans do. However, only human infants will point to things out of sheer interest.[41] Sometimes this is done to solicit a response from an adult, but more often than not the youngster is simply pointing out something interesting to be shared. No other animal does this.[42]

Copycats

In addition to joint signalling, we also copy each other. Initially, parents and babies enjoy copying each other's expressions and noises in reciprocal exchanges. Adults instinctively speak to young babies in that high-pitched, musical, gibberish language in an attempt to elicit smiles and laughter.[43] (You may have noticed that couples and pet owners can also do this.) Adults attempt to match the behaviour of the infants because babies respond to it. Sometimes, babies take the initiative and begin to spontaneously copy others around them.

These imitative behaviours are not just limited to language. Facial expressions, hand gestures, laughter and complicated actions can all be observed. Imitation signals to others that we are like them too, and we are the best imitating species on the planet. Andrew Meltzoff from the University of Washington thinks that babies really do this to establish a 'just like me' relationship with the adult.[44] They are using imitation to identify others as friend or foe. The mechanism works both ways. When adults mimic the facial expressions of infants back to them, these signals are telling the baby that this person is one of them.[45]

Before the child has reached their second birthday, they will copy all manner of behaviours. However, this is not slavish mimicry triggered automatically but rather an attempt by the infant to get into the mind of the adult. After watching an adult 'fail' to pull the end off a toy dumbbell, eighteen-month-old infants will read the true intention of the adult and complete the task they had never seen before.[46] In one study, shown in Figure 5, fourteen-month-olds watched an adult experimenter bend over and activate a light by pressing the button with her head (A). For some of the infants, the adult's hands were bound by a blanket (B).

The babies were then given the light switch to play with. Infants who saw the adult whose arms were bound (B) activated the light switch with their hand because they understood that the adult was unable to use their hands. However, if they were the ones who saw that the adult's hands were free (A), then the infants bent over and activated the button with their head too. They must have reasoned that it was important to use the head and not the hands. Infants were not simply copying the actions but rather repeating the intended goal. They had to get into the mind of the experimenter in order to work out what they wanted to achieve.[47]

Older children will copy adults' actions even when the children know the actions are pointless.[48] In one study, preschoolers watched an adult open a clear plastic box to

Figure 5:
Hands-free adult activates switch in A, whereas adult's arms are constrained in B (image courtesy of Gergely Csibra and György Gergely)

retrieve a toy. Some of the actions were necessary, such as opening a door on the front of the box, whereas other actions were irrelevant, such as lifting a rod that lay on the top. This behaviour is unique to humans. When presented with these sorts of sequences, children copied both the relevant and irrelevant actions whereas chimpanzees copied only those actions that were necessary to solve a task. The apes behaved in a way that was directed towards the goal of retrieving the reward, whereas for children, the goal was to faithfully copy the adult. Why would children over-imitate a pointless action? For the simple reason that children are more interested in fitting in socially with the adult than learning how to solve the task in the best possible way.[49]

Developmental psychologist Cristine Legare at the University of Texas at Austin, thinks that this early blind imitation observed in children has profound implications for our species. Along with her anthropologist colleague Harvey Whitehouse from Oxford University, she has been looking at the origins of human rituals.[50] Rituals are the activities that bind humans together – acts with symbolic significance that demonstrate that members of a group have shared values. All cultures have rituals for various events that are typically major transitions in life – birth, adolescence, marriage and death. These events punctuate our lives and are often associated with religious beliefs and ceremonies. The rituals themselves are typically inscrutable. There is no inherent logic to them. In that sense there are no causal laws operating, but if you don't follow the rules then the ritual is violated. There is something about carrying them out in the correct way which gives rituals their potency. Likewise, Legare has shown that four- to six-year-olds are more likely to copy a behaviour step by step that has no obvious goal compared to one that does. In doing so, the child may be beginning to understand that there are some activities that others engage in that have no purpose but must be important precisely because they serve no obvious goal.[51]

Getting into someone else's head

You cannot directly see other people's intentions, but you have to assume that they have them. This is called *mentalizing* – assuming that other people are intentional because they have minds. People are not random, but rather do things on purpose because they have goals that control their behaviour. In one study,[52] twelve-month-olds watched an experimenter look at one of two stuffed animal toys and exclaim, 'Ooh, look at the kitty!' A screen was then lowered and raised to reveal the adult holding either the kitten or the other toy. If the adult was revealed holding the other toy, the babies looked longer – they were confused by her intentions. They interpret people as doing things for a reason. If mum is looking at the sugar bowl on the table, then she is likely to pick it up but not the salt-shaker that she has not been looking at. When mum looks at and then walks over to the fridge, she does so to open it. Infants are building up an expanding repertoire of contingencies – knowing that people behave in predictable ways. When babies think that something has a mind because it appears to act as if it has purpose, they will attempt to engage in joint attention. They will even copy a robot if it appears to have a mind. By simply interacting with a baby and responding every time the baby makes a noise or an action, the robot soon becomes an intentional agent, so that babies will actively try to engage the machine and even imitate its actions.[53]

In contrast, animals do not imitate spontaneously as an attempt to initiate or engage in a social exchange. They may have the capacity for mentalizing, but invariably this is limited to situations that satisfy self-serving needs. For example, amorous male apes and monkeys will manoeuvre female partners out of the line of sight of dominant males in order to copulate surreptitiously.[54] Many animals will steal food if they believe that others cannot see the theft. All of these abilities of perspective taking are heightened when there is potential danger from a competitor. However, it is not clear that these evasive actions really involve mentalizing. I know that I can avoid the strike of a snake if I approach from out of its line of sight in much the same way that I can avoid a tumbling boulder if I coordinate my actions correctly. In neither situation do I attribute mental states. I simply observe the actions and reason about what is relevant information. To establish mentalizing, there needs to be evidence of the attribution of beliefs – states of mind that individuals hold to be true about the world in the absence of any direct evidence. If I think you have a belief, then I assume that you hold certain expectations about the world to be true.

Even then, one could attribute belief to others simply by putting yourself in their shoes. For example, we can both separately enter and exit a hotel room and I can describe what I believe you saw based on my own experience. I would reason that because we both went into the same room, you must have seen what I saw. However, that need not be true. You might have had your eyes shut or something in the room might have changed, in which case I would be mistaken. For true mentalizing ability, you need to be able to understand that someone else might hold a different view from yours and indeed be completely wrong about the true state of the world. In other words, the litmus test for real mentalizing is the understanding that someone can hold a *false belief*.

Consider the following test. If I were to show you a confectionery box with 'M&Ms' written on it and ask you what is inside, then in all likelihood you would answer 'M&Ms'. However, if I open it up to reveal pencils, then you should be a little surprised and possibly a little annoyed because you expected a chocolate treat. If I ask you what you originally thought was in the box, you would say 'M&Ms' because you understand that you had a false belief. This may seem trivially easy, but most three-year-olds give the wrong answer and claim that they thought there were pencils in the box.[55] It's as if they have completely rewritten history to fit with what they now know to be true. They do not understand that they held a false belief. Understanding that someone can be mistaken is part of a capability called *theory of mind* and children operate with an increasingly complex set of assumptions about the minds of others.

If three-year-olds do not understand that they were mistaken, then it is not too surprising that they are unable to attribute false beliefs to others. If I ask you what someone else will answer when posed the same question about what's in the box, then you understand that they, too, should answer 'M&Ms'. You can see things from their perspective and understand that they will also have a false belief. Again, three-year-olds give the wrong answer and say pencils. It's as if they cannot easily take another person's perspective.

When young children act in this self-centred view they are said to be *egocentric* because they view the world exclusively from their own perspective. If you show young children a model layout on a tabletop of a mountain range with different landmarks and buildings and then ask them to select a photograph that corresponds to the view they can see, three-year-olds correctly choose the one that matches their own perspective. However, when asked to choose the picture that corresponds to the view that someone else standing on the opposite side of the table can see, they typically choose again the photograph that matches their own.[56]

Young children cannot easily formulate a mental picture of what it is like to see the world from someone else's viewpoint. The classic demonstration of this false belief perspective involves two dolls, Sally and Anne.[57] In the Sally–Anne task, Sally has a marble that she puts in a toy chest before saying goodbye to Anne and leaving the house. Whilst she is out, Anne moves the marble from the toy chest and places it in the cupboard under the sink in the kitchen. The child is asked where Sally will look for her marble. Adults easily know that Sally will look in the original location. After all, she does not know that Anne moved the marble and she isn't psychic! Again, three-year-olds fail the test and say Sally will look in the cupboard in the kitchen, under the sink.

Why does it take so long for young children to understand that others can be mistaken? After all, infants understand that adults behave purposefully when watching their actions. One explanation is that young children do not yet understand that others have minds that can harbour false beliefs. Another explanation is that these tests require individuals to make a response that runs counter to what they know to be true. They have to actively ignore the true state of the world. If the task requirements are changed so that the need to respond is taken away, then a different picture emerges. One study examining the looking behaviour of infants reveals that they will look longer when Sally, who should hold a false belief, goes to the correct location as if she knew that her marble had been moved to a new location.[58] Sally's psychic ability creates a violation of expectancy in the infants, so that they are surprised.

Appreciating that others can have false beliefs appears to be uniquely human, as there is no compelling evidence that other animals can acquire this aspect of theory of mind. As noted earlier, they can consider another's perspective, which is how animals learn to deceive or pay attention to potential competitors; but they do not reliably pass tasks that require understanding that another holds a false belief. When tested on a similar non-verbal version of the Sally–Anne task, apes fail when required to make a choice by looking for food in one of two locations; but like human infants, they seem to register some indecision by looking longer or backwards and forwards between locations when there has been a surreptitious switch of target from one hiding place to the next.[59] Together, the looking measure suggests that there is some rudimentary knowledge about mentalizing present in both apes and young infants. However, only in humans does that understanding develop into a full theory of mind that we observe in typical four-year-olds.

Working out what others know is not always as trivially easy as the Sally–Anne task. Consider more complicated plots with more characters and more changes of events. When someone says 'I know that she knows that he knows', then they are applying multiple theories of mind. Keeping track of who knows what can easily become more difficult with each layer of plot added. Even then, you have to pay attention because if you miss a key step or forget who did what, you get it wrong.

Add to this the trouble with knowledge. When we know something is true, it is harder to ignore the content of our own minds when attributing a false belief to others.[60] We have to actively suppress our own knowledge in order to correctly identify the state of mind in another. As we shall see later, in Chapter 4, deciding *not to do something* requires actively *doing something,* which may be compromised in young children and absent in most other animals. So even adults who pass Sally–Anne tasks take longer to correctly attribute false beliefs to others. They are also much slower to solve false-belief situations when you give them a second task that occupies their own minds. It takes effort to think carefully about what others are thinking. Also, it is not clear that adults always employ a theory of mind during most social interactions.[61] When you open the door for someone, do you really try to work out what his or her intentions are or do you mindlessly execute an action out of habit? Just because we can generate a theory of mind does not mean that we always do.

New York psychologist Lawrence Hirschfeld argues that while mentalizing through a theory of mind might be one way of predicting and interpreting someone's behaviour, a better strategy, which is more accurate and efficient, is to make certain assumptions about the situation. In many of our interactions with others, we do not try to infer what is on their mind at all. Holding doors open for others, for example, is a mindless act, as are many of our social interactions.[62] This is because humans may not be that good at attributing the correct mental state to others in the first place but they are better at reading what is normal behaviour in different contexts. Rather, we learn to apply a *theory of society* interpretation to the motivation of individuals – what people typically do in a particular situation. This would be based on learning about different members of the group as defined by the different categories they occupy, such as age and gender.

We operate with stereotypes, which lead us to assume that people will behave in certain predictable ways based on past experiences. This may actually be the default strategy for reasoning about other minds. In other words, it is when people do something we regard as unusual that triggers our mentalizing, as in, 'What the hell were they thinking?' This is when our false-belief reasoning is switched on, in an attempt to rationalize another's actions. The idea that children learn about such exceptions to normality is supported by studies that show they are more likely to seek an explanation when they encounter variability in another's behaviour.[63] They are also more interested in inconsistent outcomes – like detectives trying to solve puzzling behaviour.[64] They seem driven to try to understand the social world around them by making sense of people as predictable agents. Children need to learn what is typical for certain individuals as opposed to what is typical for most people.

How we make up our minds

Babies are clearly not just little adults, so what sort of creature are they? They are not blank slates: they are born with a brain that is already prepared for learning about the world. They have an instinct to learn. The development of the mind through learning must be the interaction between brain and the environment, shaped by mechanisms that have evolved to make sense of the world. But how much is built in by evolution and how much of it comes from experience?

As complicated animals, we engage in complex levels of analysis of the world. We have raw sensations streaming in through the senses that have to be organized into meaningful patterns that reflect information and structure in the environment. It would be a blooming buzzing confusion if it were not for the fact that we have some rules about how to make sense of our senses. These are the perceptual processes in the brain that detect and generate patterns. However, perceptions are only of use if they can be stored and recruited for future reference in order to plan behaviours. This is the job of cognition or thought. We can think about what we have learned and apply that knowledge to predict what to do next in a situation.

For young children, much of that world is a social one because they are so reliant on others for their survival. In the same way that we are adapted to understand some features of the physical environment, we also seem to be adapted to learning about others. Rudimentary social systems need to be fine-tuned or switched on by experience so that we can begin to understand people.

Some animals can also read other's behaviours, but only when it is in their interest to benefit. Most animals are selfish, with little concern for others. In contrast, during the first year, a human baby's social interactions with adults are rich and numerous but it is not clear that infants fully understand that the adult has a mind of their own yet. Without language, it is not clear that we could ever know what a baby is thinking about others. Maybe they are just like meerkats who automatically follow another's direction of attention. However, as they grow, babies become more interactive with the world around them and seek out the attention of others. They may not have language by their first birthday but they are already communicating and reading non-verbal signals. They have gestures, squeal, blow raspberries, pull faces, protest, throw toys, point out things of interest, show fear or happiness and, of course, cry. Not only can they signal to adults what's on their mind or at least when they are happy or unhappy, but they are beginning to understand that adults have minds too. When we can understand the minds of others, we can predict what they will do in the future. That is an enormous advantage when making sense of those around us.

Knowing what someone will do by reading their mind is one of the most powerful things our brains can do. When you know what someone else is thinking, you can manipulate and out-manoeuvre them for strategic advantage, just like Machiavelli. Even when you are not in competition with others, you still need the ability to know what they are thinking. Before language evolved, it would have been critical to understand what was on someone else's mind so that you both could share the same perspective. You have to be able to put yourself in someone else's situation in order to understand their intentions.

From sensation to culture, social mechanisms form a multi-layered system that is embedded in the newborn brain through natural selection but ultimately shaped and operated within a cultural environment. They are the tools that bind us together in a shared world. But there are other mechanisms that bind us together – we do more than share attention and interests, we also share emotions. From the very beginning, we are immersed in an emotional world where others make us feel happy or make us feel sad. The drive to have children may come from our selfish genes, but these genes also build the mechanisms that fuel our behaviour by providing feelings. Who we become is largely shaped by the emotions that motivate us, but these drives can be shaped by early experiences that leave a surprising legacy.

CHAPTER 3

Getting Under Your Skin

There was a time when it was acceptable to stare at individuals who, through the misfortune of the lottery of life, had been dealt a bum hand of cards when it came to their genes. Regarded as 'freaks of Nature', they came in all sorts of shapes and sizes – the victims of genetic abnormalities. These included dwarves and giants, people without limbs, bearded women, albinos and, most famously of all, the severely deformed Joseph Merrick, also known as *The Elephant Man* because of the massive tumours that disfigured his face and body.[1] Although Merrick went on to lead a celebrity life, most of these people ended up in travelling circuses or freak shows where the public would pay to simply gawk at them.

In an attempt to understand such unfortunates, a widely held view at the time was that the birth defect was caused by some frightful event that traumatized the mother when she was pregnant. This idea, known as *maternal impression,* is thousands of years old and reflected a common belief that there was a correspondence between the nature of the birth defect and the supposed shock. A mother being accidentally burned during pregnancy may cause a patch of discoloured skin on the baby. Cleft palates or harelips occurred because a leaping hare had surprised the mother. Or, more commonly, the pregnant woman was so frightened by the sight of some deformity on another person that her unborn baby would be afflicted by the same defect. In the case of Joseph Merrick, it was claimed that a rogue fairground elephant startled his pregnant mother.[2] These ludicrous ideas are consistent with magical thinking – the idea that there is a causal link between two events that are similar in appearance rather than an unrelated coincidence.

Although magical thinking has been largely abandoned in the West since the nineteenth century, maternal impression is still widely believed in many parts of the world today.[3] Some countries have rituals, talismans and customs to ward off harm to protect the unborn child. In India, pregnant women avoid certain individuals such as barren women who may affect their foetus by casting the 'evil eye'.[4] While it may seem absurd that frightening a pregnant woman would have a permanent effect on her offspring, recent findings suggest that we may have been a little too hasty to dismiss maternal impression, or at least the susceptibility of unborn children to traumatic external events.

In this chapter we examine the possibility that early domestic environments not only shape what we learn, but also how we respond emotionally in terms of temperament. Temperament refers to the individual differences people have in their emotional responses. Some of us are more anxious whereas others are more outgoing. Some are more aggressive and others are more fearful. From the very beginning, babies differ in temperamental styles in that some cry more easily or startle suddenly whereas others are more laid-back and placid. Individually, we tend to be more like our parents when it comes to our emotional dispositions, which indicates this dimension of personality has a genetic contribution. However, early environments can also shape the development of temperament in ways that shape who we become as adults, and how well we adapt to domestication.

The day the world stood still

I can still vividly recall it as if it were yesterday. Those of a certain age will remember exactly where they were on that fateful day in 2001. It was a September afternoon in the UK but a bright, sunny morning in New York with crisp blue skies. Colleagues knew that I had a television in my office and had come in to watch the terrifying news unfolding. Two planes had been flown into the World Trade Center and now there was dense smoke billowing out of both. People were jumping to their deaths. If you saw the footage, then you, like me, will probably still have those events emblazoned on your memory as the world changed for ever.

For some, these recollections have become *flashbulb memories*, as if the scene were lit up in harsh lighting to capture everything – even trivial details of little relevance. Our memories can be supercharged with detail when we experience something terrifying. This is because we become more alert and attentive, on the lookout for danger as our hippocampus, the seahorse-shaped repository for long-term memories in each of the temporal lobes, receives input from the amygdala – a structure the size of an almond, also in each temporal lobe, that is active when you laugh, cry and scream in terror.[5] They also don't let you forget.

Experiences that eventually become memories start out as patterns of neural firing or traces that come flooding into the brain. Raw sensory input is interpreted into representations and given meaning. This in turn updates and changes the knowledge we have about the world by forming memories. Whether details become consolidated into the memory stores of the hippocampus depends on filtering mechanisms that are regulated by the action of neurotransmitters released by the amygdala during surprising, arousing or rewarding events. The neurotransmitters are the molecules that trigger activity in the connecting gaps between neurons. Flashbulb memories stimulate the amygdala to invigorate the activity of the hippocampus, thereby enhancing the memory trace for those events that move us the most.[6] As the world watched in helpless shock, a generation would never forget what they saw. But even some from the next generation of unborn babies were left with the legacy of that terrible day.

Post-traumatic stress disorder (PTSD) is an anxiety condition that appears weeks after traumatic events such as rape, battle and other acts of violence. It is characterized by recurrent dreams, flashbacks and flashbulb memories, as if the victim is haunted by the past. After witnessing 9/11, one in five New York residents who lived closest to the World Trade Center suffered from PTSD. Rachel Yehuda, a New York psychiatrist, followed up a sample of pregnant women from this group. She found that these women had abnormal levels of cortisol in their saliva – a hormone that is released as a natural response to stress but depleted in individuals with PTSD.[7] Different hormones and neurotransmitters form part of an elaborate signalling system that the brain uses to activate different functions. Some have general effects whereas others seem to be more specific in the roles they play.

The depleted levels of cortisol in the chronically stressed mothers were to be expected. But what was unexpected was the plight of their unborn children. One year after the attack, infants born to the mothers who had developed PTSD also had abnormal levels of cortisol compared to babies of other mothers who did not develop the disorder after witnessing 9/11. Vulnerable mothers had passed something on to their children. As Yehuda put it, children of PTSD victims bore 'the scar without the wound'.[8]

It is well known from various disease models that events early in development can have consequences later in life. There is a whole category of substances known as *teratogens* (literally, 'monster makers') that, if the pregnant mother is exposed to them, can result in birth defects. Various drugs, both legal and illegal, as well as environmental toxins such as radiation or mercury can damage the unborn child. However, some diseases resulting from harmful substances take decades to manifest. My own father-in-law died from mesothelioma, a rare form of lung cancer that was probably caused by exposure to asbestos when he was growing up as a child in South Africa. Toxins that enter our bodies can alter the functions of our cells but lie dormant for years. Over a lifespan we may replenish our cells many times, but each reproduction of the cells can carry genetic time bombs that lie in wait for the right circumstances to kill us. Physical substances like asbestos from the environment are obvious candidates as being poisonous to our systems, but what about exposure to psychological toxins? How can our mind's reaction to non-physical events, such as watching something horrific, produce long-term consequences? How could a mother's stress in response to 9/11 cross over to the next generation? What could she possibly pass on to her unborn child?

Jerry Kagan, a Harvard developmental psychologist, reckons that around one in eight babies are born with temperaments that make them highly irritable, which is due to their over-reactive limbic systems. They startle easily and respond excessively to sudden noises.[9] The limbic system mobilizes the body for action and its circuitry includes the amygdala. It triggers a cascade of hormones and neurotransmitters that prepare the body to respond to threat. Reactivity of the limbic system is a heritable trait meaning that it can be passed on to the child in the genes they inherit.[10] These are the highly-strung children who find uncertainty and strange situations upsetting. Depending on how they react to sudden sounds as a four-month-old baby, you can even predict personality many years later.[11] Reactivity is like a disposition, which makes some of us twitchy, but others are born more laid-back and chilled. Maybe mothers who developed PTSD after 9/11 gave birth to babies with a nervous nature because of their genes.

Yehuda thinks not. She found that the lowered cortisol effect was only present for those mothers who were in the third trimester of their pregnancy, so it could not just be the genes working alone. There seems to be a critical period when exposure to stress alters the child's development. To begin to understand how such a maternal impression restricted to a window of vulnerability could possibly happen, we need to look at the history of difficult childhoods and the way that they affect how we respond to stress as adults.

War child

World War II disrupted normal life for thousands of families. In Europe, many children were separated from parents by the turmoil and ended up in institutions. Even though they were generally cared for, many of them grew up into socially impaired and delinquent teenagers. To explain this, John Bowlby, a British psychiatrist, proposed that these children had missed out on a critical phase in development that he called *attachment*.[12] Bowlby believed that attachment was an evolutionary adaptive strategy to form a secure, nurturing bond between the mother and her infant. This early experience not only protects the vulnerable child, but also provides the necessary foundation for coping mechanisms to deal with problems later in life. Without this early secure attachment, the child would grow up psychologically impaired.

Bowlby was inspired by the ornithological work of Konrad Lorenz, who had shown that many bird species form a close-knit bond between mother and chicks.[13] This attachment begins with imprinting, where the young chicks will pay special attention to and follow the first moving thing they see. Famously, Lorenz demonstrated that he could make baby goslings imprint on him by incubating the eggs and hand-rearing the chicks when they hatched. In the wild, imprinting was critical for survival by maintaining the proximity of the chicks to the hen, which is why the chicks would imprint to the first moving thing, usually the mother. Investigation of the chick brain revealed that it is innately wired to follow some shapes more than others and that chicks quickly learn the distinct features of their own mother, to tell her apart from others.

Human infants also pay special attention to face patterns at birth and very quickly learn their mother's face.[14] However, primate, and in particular human, early social attachment is unlikely to be as rigid as bird imprinting. Whereas the need to imprint in birds has to be satisfied fairly quickly, primates can take a bit longer to learn to know each other. Another important difference between birds and babies is that humans are not up and running about for at least a year. Whenever the human infant needs their mother, they simply have to cry, which will soon send most mothers scurrying to their infant's side. A distressed infant's cry is one of the most painful things to hear (which explains why crying babies on aeroplanes can be so upsetting for everyone around them). This 'biological siren' ensures that babies and mothers are never that far apart.[15] Infants from around six months of age show separation anxiety when physically separated from their mother, a state characterized by tears and stress as signalled by the rise in cortisol levels in both the infant and mother. These levels eventually return to normal when baby and mother are reunited.[16]

With time, both mother and baby learn to tolerate further episodes of separation, but the mother remains a secure base from which the toddler can explore their surroundings safely. Imagine Bowlby's securely attached toddlers as baseball or cricket players: they feel secure when they are touching the bases or while behind their creases, but become increasingly anxious and insecure as they step farther and farther away from them. Without secure early attachment, Bowlby argued that children would never learn to explore novel situations and develop appropriate coping strategies. They would also fail to become properly domesticated, which was why he believed that children separated from their nurturing parents during the war grew up to become delinquent teenagers.

The lost children

Inspired by Bowlby's work on social attachment and later psychological abnormality, Harry Harlow in the US set out to test an alternative explanation for the long-term effects of deprived childhoods.[17] Maybe children were simply not looked after or given adequate nutrition if they were raised in institutions. If you gave them food and warmth, they should be fine. To test this, he conducted an infamous series of studies where he raised baby rhesus monkeys in isolation for differing amounts of time. Although these infant monkeys were well fed and kept in warm, safe environments, they were left alone. This social isolation had profound effects on their development. Monkeys with no social contact as infants developed a variety of abnormal behaviours as adults. They compulsively rocked back and forth while biting themselves, and when they were finally introduced to other monkeys, they avoided them entirely. When the females from this group reached maturity, they were artificially inseminated to become mothers, but they ignored, rejected and sometimes even killed their own offspring.

Harlow discovered that it was not just the amount of time that animals spent in isolation that was critical, but also when they were separated. Those born into isolation were at risk if they spent longer than the first six months without the company of their mother. In comparison, monkeys who were isolated only after the first six months of normal maternal rearing did not develop abnormal behaviour, indicating that the first six months was a particularly sensitive period. Bowlby had originally thought that the primary reason for attachment was to ensure that biological needs for food, safety and warmth were satisfied, but Harlow proved that he was only partly correct – monkeys also needed social interaction from the very beginning.

It turns out that human social development, like that of the monkey, is also shaped by a similar sensitive period of socialization. Back in 1990, following the collapse of Nicolae Ceauşescu's dictatorship, the world discovered thousands of Romanian children abandoned in orphanages. Ceauşescu had outlawed family planning in an attempt to force women to have more children to increase Romania's dwindling population. The trouble was that families were unable to support these children and so they were abandoned in the orphanages.

On average, there was only one caregiver for every thirty babies, so there was little social interaction and none of the cuddling or intimacy that you would find in a normal, caring environment. The babies were left to lie in their own faeces, fed from bottles strapped to their cots and hosed down with cold water when the smell became unbearable. When these children were rescued, many of them were fostered out to good homes in the West. Sir Michael Rutter, a British psychiatrist, studied just over one hundred of these orphans who were less than two years of age to see how their early experiences would shape their development.[18]

On arrival, the orphans were all malnourished and scored low on psychological tests of mental well-being and social interaction. That was to be expected. As time passed, they recovered much of this lost ground in comparison to other adopted children of the same age who had not been raised in the Romanian orphanages. By four years of age, most of this impairment had gone. Their IQs were still below average in comparison to other four-year-olds, but within the normal range that could be expected. However, it soon became apparent that not all was back on track.

Children who had spent longer than six months in the orphanages were failing to catch up with their matched group. Only the children who were rescued before they were six months old went on to a full recovery. The children were followed up again at six, eleven and fifteen years of age. Again as a group they fared much better than expected, given such a horrific start in life, but problems started to appear. Those who had spent the longest time in the orphanage were beginning to show disturbed hyperactive behaviour and difficulties in forming relationships. Just like Harlow's monkeys, social interaction during that first year was crucial for normal development. To understand what is so important about having someone around to look after you and not just to give you food and warmth, we have to consider what upsets babies.

Why not knowing is stressful

Have you ever waited for someone to call you with important news? Maybe it was an exam result, the outcome of a job interview or, worse, a call from the hospital. The reason that waiting for important news produces anxiety is that brains are pattern detectors that have evolved to seek out regularities in life; not being able to predict what will happen next is therefore upsetting. We can brace ourselves for important events, but it is stressful to maintain that level of preparedness for a long time. The stress comes from high levels of arousal – a state of readiness and expectation. Just like the US Army, when we face a threat, we enter a defence-readiness condition (DEFCON). When that threat is at a peak, it's like being at DEFCON 1. This is why we jump at the slightest sound, because we are in a state of heightened alertness. It is not until we can stand down our defences that we can relax.

Even though we may not be actively dealing with a threat, nevertheless the uncertainty of threat still makes the situation stressful. In fact, our brains are not very good at dealing with random events, which is why we tend to see structure and order everywhere. That is why when you are in the woods late at night or an old spooky house, every noise sounds like a threat. Adults start to see patterns in random noise when you remove their ability to control outcomes or remind them of times when they were helpless.[19]

This lack of control is not only psychologically distressing, but it also affects how our bodies respond. Even our tolerance to pain is reduced. Adults can withstand much more painful electric shocks if they think they can stop the punishment at any point in comparison to those who do not think they have this option.[20] Believing you can stop the pain whenever you want means that you can tolerate more. However, when faced with unpredictable and uncontrollable shock, both animals and humans develop both psychological and physiological illness.

This need for control and predictability is present from the very start. Babies prefer regularity and predictability, which is why they startle to sudden unpredictable noises, lights or movement. In fact, there is a reflex controlled by the brain stem – the most primitive part of the brain that controls vital functions – known as the *startle reflex*, that jolts the child to attention. If a newborn does not startle, there is a chance that they have some form of damage to their nervous system. This need for predictability forms the basis for contingent behaviour where the baby starts to learn how they are synchronized with others. Such a sensitivity to external events means that a nurturing domestic environment is one that is going to be predictable and less threatening – attributes that can be controlled by caregivers.

Infants thrive on predictable contingent behaviour, but the flipside is that they find unpredictable or non-contingent events upsetting, especially when they involve their mother. When mothers are depressed, they often have flattened emotions and so the quality of their interaction with their infants is impoverished.[21] Other depressed mothers, rather than being sad and dull, over-compensate in an animated, exaggerated form of communication, which can be equally distressing to the infant because it is not contingent with their own efforts at communication. Early experiences like this, where the baby's needs for contingent responses are not met, can lead to social and cognitive difficulties many years later.

Other people provide reassurance in an uncertain world. The stress of uncertainty is reduced if there is an adult around, so our brain benefits not only from the wisdom of others but their presence as well. As the saying goes, a problem shared is a problem halved because there is strength in numbers. If you think about it, the world is full of surprises for the young infant and development must include discovering what is going to happen next. With knowledge and experience, the world becomes more predictable. That understanding takes time to acquire, but until then, adults provide protection and reassurance, which is why babies cry if there is uncertainty because it is how they signal to the adult to resolve the situation.

Taken together, these studies indicate that extreme early environments can have long-term effects on developing monkeys and humans. It would seem that primates need some form of contact from the very start, especially in environments that are particularly threatening or socially vacant. However, it is not just the deprivation of not having others around, but not having others around who are reliable. How do such stressful environments shape who we become and what role do others play in our reaction to stress?

Learning to fight or flee

To understand how aversive unpredictable environments affect growing brains, we need to understand the normal response mechanisms to stress. When faced with a threat, we can either stand up to it or run away. There is a rapid *fight-or-flight* response where we get that sudden emotional rush that requires us to mobilize as quickly as possible that is triggered by activity in the limbic system of the brain. This preparedness is achieved by a system called the *hypothalamic-pituitary-adrenal* (HPA) *axis.*

Following exposure to stress, the hypothalamus releases two hormones, corticotropin-releasing hormone (CRH) and arginine vasopressin (AVP), that stimulate the nearby pituitary gland to release adrenocorticotrophic hormone (ACTH) into the blood stream. ACTH targets the adrenal glands that sit atop the kidneys way down in the guts to release adrenaline, noradrenaline and cortisol. The balance of adrenaline and noradrenaline regulates the autonomic nervous system (ANS), which in turn increases breathing, heart rate, sweating and pupil dilation, and shuts down digestion. After all, you don't have time to chew the cud when you are about to do battle. If you have ever felt butterflies in your stomach before going on stage, that was your ANS operating. Cortisol works by increasing the concentration of glucose in the bloodstream, thereby making more fuel available for muscles. All of this activity is fine when there is a real threat that needs to be

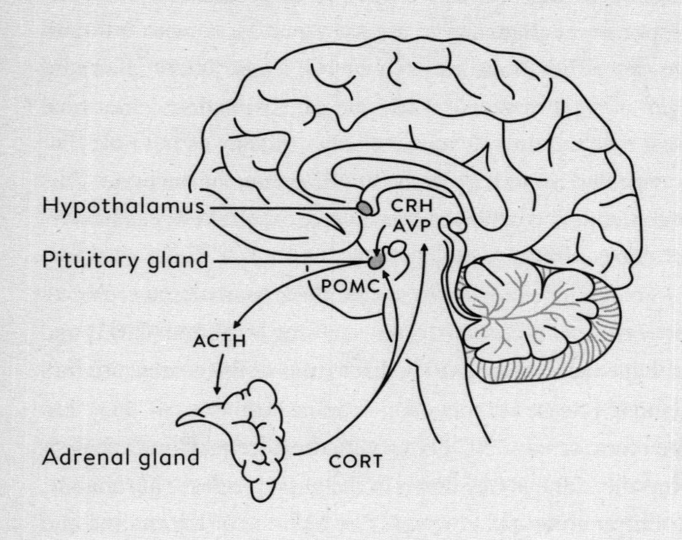

Figure 6: The hypothalamic-pituitary-adrenal (HPA) axis

dealt with immediately. However, the fight-or-flight response has to be wielded appropriately and used in moderation.

Maintaining high levels of stress over long periods leads to chronic impairment in our ability to cope with life's ups and downs. It is like keeping your foot on the accelerator pedal, revving the engine, and it will eventually cause damage to the HPA axis mechanisms and subsequent illness and impairment of your immune system. Chronic stress has also been linked to psychiatric disorders such as depression, with most individuals suffering from major depression having increased HPA activity.[22] So, to keep body and mind in a healthy state, you need to be able to regulate your stress response. Part of this regulation is provided by the hippocampus. Within the hippocampus, there are glucocorticoid receptors (GR) that monitor levels of glucose and cortisol in the bloodstream. When levels of circulating glucose and cortisol reach a critical level, the hippocampus signals the hypothalamus to shut down the HPA process in the same way that a thermostat on a heater regulates temperature. If a thermostat is faulty, the house freezes or overheats. Likewise, if the HPA is disrupted, either you do not respond adequately to stress or you overreact.

Children raised in abusive households suffer not only from episodes of violence and harm but also from the unpredictability of when the next abuse will happen. Unpredictability is corrosive to coping, as we are not able to relax but must maintain our stress response in a state of high alert. This will produce long-term disruption of the HPA system, which can have consequences many years later. This may be one reason why PTSD sufferers have abnormal patterns of circulating cortisol, because their HPA remains on high alert and is unable to relax.[23] In a study reminiscent of Bowlby's original work, Finnish scientists followed up 282 children evacuated during World War II to test the effects of separation from parents on their stress responses decades later. Those separated from their parents during the war as young children had higher cortisol reactivity to stress tests sixty years after the early separation, indicating that the physiology of their HPA system had been altered permanently by this experience.[24] The older the child was at the time of the evacuation, the more resilient they were and the less disruption to their HPA system as adults.

Even before you are born, stress can alter the functioning of the HPA axis. Female rhesus monkeys in the later stages of pregnancy were taken from their cages and exposed to unpredictable, loud, stress-inducing noise. After giving birth, not only did these young mothers have disrupted HPA responses, but so did their offspring, in comparison to other mothers who had not been stressed during pregnancy or their infants.[25] In the same way, experiencing a terrifying, unpredictable event like the attack on the World Trade Center in which no one knew what was going on, some pregnant mothers may have inadvertently passed on a legacy of fear to their unborn children.

Once born, the long-lasting effects of early exposure to stressful households have been shown to alter the way babies respond to aggression even when they are not awake.[26] Infants between six and twelve months of age had their brains scanned when they were asleep inside an fMRI scanner. They were played nonsense sentences spoken in very angry, mildly angry, happy and neutral tones of voice by a male adult. Even though they were unconscious, those babies from households where there were high levels of conflict showed greater reactivity to the very angry voice in the ACC, caudate, thalamus and hypothalamus – all brain regions of the HPA system. Already their stress response had become sensitized to the presence of aggression.

The HPA system is also altered in animals that become domesticated. As we saw earlier, domestication produces changes in behaviour and the brain. Domesticated animals are less fearful, less aggressive and have elevated levels of serotonin[27] – a neurotransmitter associated with prosocial activity.[28] Normally wild fox pups become fearful of humans at around forty-five days and are less likely to explore their environments as their natural fight-or-flight response kicks in. In contrast, this fearful response is not observed in domesticated pups of the same age and they continue to explore their environments. For domesticated foxes, the period of socialization is significantly longer and play activity extends into adulthood.[29]

Don't feel nervous, feel excited!

The relationship between body and mind is critical to understanding emotion. One of the first accounts of this relationship was William James's proposal that emotions were produced by the body's response to a stressful encounter.[30] When we see a bear, our fight-or-flight response immediately kicks in to deal with the threat and only afterwards do we feel the emotion of fear. That's the way it should be, as a good evolutionary strategy, because it is better to act first and ask questions later when you are in potential danger. James argued that we needed to react before we had time to ponder the situation. You don't want to be sitting around considering how you feel about the bear.

Most of us rarely encounter bears in the modern world, but we have all had that act-now, think-later experience. Maybe it was a sudden fright when someone jumped out at you or possibly an unexpected threat. Our heart rate and breathing surges as adrenaline pumps around our body in preparation. Road rage is a classic aggression scenario triggered by a perceived threat before we have had time to evaluate the actual threat.

James's account of emotions following responses failed to take into consideration situations where the body responds more slowly to stressful situations than our thought processes.[31] Also, people are not always sensitive to the changes in their body in stressful situations. Sometimes emotions can precede bodily changes, which is why we can feel embarrassment before we blush. Maybe you burped in public accidentally, looked around at others and then felt your cheeks burning bright red with heat as the emotional significance of your *faux pas* sunk in. The thought was almost immediate but the change in blood flow took longer. So which is it? Does fleeing cause fear or do we run away because we are frightened?

The answer is both. In some situations, the need to respond as fast as possible trumps the need to think (the sudden bear attack), whereas in others we need to consider the situation and respond accordingly (blushing in public). However, in both situations, experience and expectations play a role. If we know that the bear is actually stuffed, then we are less likely to be frightened. If we are among family when we burp, we do not feel so socially awkward.

As these different examples reveal, there are fast and slow pathways to emotion that depend on the circumstances and how we interpret the situation.[32] Our emotions are also largely influenced by others. In a classic study of the importance of social context,[33] naïve subjects were given an injection of adrenaline and told they were receiving vitamins that would boost performance on a visual test. This was all a sham to get at the real purpose of the study – how do those around us influence emotional experiences? Some of the participants were correctly informed that the injection would make their hands tremble, give them a flushed face and increase their heart rate. Others were told incorrect symptoms of a mild headache and itching skin.

While the participants sat around in the waiting room, they were asked to fill out mood questionnaires. Seated among them was an experimenter who acted in one of two ways. This confederate had not been injected with adrenaline but behaved either negatively, complaining about the study, or positively, by saying how much they were enjoying the experience and acting up playfully.

Meanwhile, in the real participants, the adrenaline triggered their HPA axis and produced the bodily symptoms associated with the fight-or-flight response. Suddenly they had these sensations, but what did they make of them? Those who had been warned correctly about the effects of adrenaline interpreted their sensations correctly ('I'm feeling a little revved up because of the shot'). But those who did not expect the increased heart rate and tremors were in a state of ignorance and needed to make sense of the signals their bodies were sending them. This is where others play a critical role. The emotions experienced by the naïve participants depended on the influence of the stooge in the room. Those seated with the playful experimenter rated their mood much more positive compared to those seated with the irritated experimenter. They were using the social context of others to interpret their own bodily sensations. Whether we are enjoying a rock concert, a football game or a day at the funfair, our emotional experience depends heavily on how others respond.

The importance of interpretation explains why some of us feel anxious and some of us feel excited. We learn to interpret situations based on experiences that we accumulate over our lifetime. This is why children raised in an environment where there is excessive conflict come to expect this as normal. If there is one thing that is predictable in conflict households, it is anger. When there is anger, violence soon follows, which is why abused children tend to see anger earlier in faces and interpret faces as being angrier whereas they show no higher sensitivity for other emotional expressions. Having a bias for interpreting anger means that children can be prepared for fight-or-flight.

Knowing this enables us to change the way troubled teenagers behave. Colleagues in my department at Bristol produced a series of computer-generated faces made up from morphed real faces that varied on a continuum from happy through neutral to anger.[34] The teenagers, most already with criminal convictions and attending a programme for high-risk repeat offenders, saw the ambiguous faces as more aggressive. However, in a clever twist, half of the teenagers were given false feedback on a task where they had to judge the expression, which eventually shifted their bias away from angry faces. In other words, after training, they were much more likely to see ambiguous faces as happy and happy faces as even happier.

The psychologists were able to shift the teenagers' perception to a more positive interpretation. More remarkably, the effect was long lasting and altered their behaviour in general. The teenagers kept diaries and were evaluated by staff who were unaware of which condition each teenager had been in. After only two weeks, those teenagers who had their anger bias shifted were happier, less aggressive and involved in less conflict incidents as rated by the staff.

Domestic violence

We all need someone from the very start. This imperative to have someone in your life explains the paradox of children's attachment to abusive parents and why domestic violence can persist. According to the UK's National Society for the Prevention of Cruelty to Children statistics published in 2012, one in four young adults were severely maltreated as children. You would think that we have evolved brains that learn to avoid danger, yet when social workers, doctors, or police officers attempt to rescue these victims from an abusive situation, the child will often lie to protect the parents. Harry Harlow also demonstrated similar phenomena in his rearing studies, when frightened infant rhesus monkeys would cling to a surrogate mother made of wire, cloth and a plastic head. Even when they were punished for this attachment with an aversive puff of air, they would still hang on for dear life. How can we understand such strange love?

Regina Sullivan, a neuroscientist who studies the neurobiological basis of attachment, believes an answer might be found by looking at rat pups.[35] Rats are smart and can quickly learn what is painful. They can learn to associate an odour with a painful shock. Surprisingly, the brain area responsible for fear and avoidance learning is turned off by the presence of the mother. Even though rat pups can associate a smell with a painful shock, they do not avoid the odour when the mother is present and will in fact approach the smell associated with punishment. Somehow the presence of the mother switches avoidance into approach behaviour in painful situations. The explanation for this masochistic behaviour is that learning about painful situations requires the activity of the rat's equivalent of the stress hormone, corticosterone, but the presence of the mother turns this off in the young pups in the nest.

Outside the nest, when they are older, exploratory rats will avoid potential dangers but they do this by returning to the nest for comfort and safety.[36] This response is *social buffering* and we see it in humans faced with stressful situations where the presence of a loved one makes the experience more bearable. Even having the photograph of a loved one is sufficient to alleviate pain.[37] The problem arises when that loved one is also the source of pain and danger. When rats return to their nest, their corticosterone mechanisms are switched off and they forget what a monster their mother can be. So unpredictable environments are stressful but less stressful than consistently abusive situations. For some, uncertainty of the future is worse than the predictability of the current situation, albeit abusive, which is the origin of the saying 'Better the devil you know'.

Clearly early domestic violence can leave a lasting impression, but not everyone responds to adversity in the same way and not everyone develops stress-related illness. Not everyone stays in an abusive situation. Given our understanding of stress as a biological phenomenon, how is it that individuals can respond to it so differently?

Two peas in a pod

I have a collection of rare postcards from the sideshow era that I described in the opening to this chapter. They fascinate me since they are a reminder of how social history and attitudes can change so dramatically. One of the cards is a rare photograph of Daisy and Violet Hilton as babies. Daisy and Violet were Siamese twins – two identical sisters conjoined at the hips. They were born in 1908 in Brighton and immediately rejected by their unmarried mother, who thought they were a curse from God for being born out of wedlock. Daisy and Violet were adopted by their midwife and raised to be talented musicians who went on to achieve fame and even appeared in the movies, most notably Tod Browning's infamous production of *Freaks* in 1932.

Identical twins occur when a fertilized egg splits in two soon after conception. In the rare cases of conjoined twins, that separation is incomplete. Identical twins share all their genes whereas non-identical twins, who come from two separate fertilized eggs, share only half of their genes. Like Tweedledum and Tweedledee from *Alice in Wonderland*, identical twins look the same, behave the same and often think the same thoughts. There is even a popular myth that twins are telepathically connected and read each other's minds.

Studying twins is important for working out the roles of genes and environment in shaping the course of development. Like Daisy and Violet, twins are sometimes adopted, but unlike conjoined twins, they can be fostered out to different households. By comparing twins, identical and non-identical, raised in the same or different households, you can estimate how similar they are and then work out the relative contribution of genes and the relative contribution of the environment.

These adoption studies show that identical twins raised separately are more similar than non-identical twins raised by different families. This proves that aspects of personality and intelligence must be heritable. But identical twins are not identical. Even as conjoined twins, Daisy and Violet had marked differences in personality and allegedly even different sexual orientations, but they were hardly the same person.[38] When it comes to personality and intelligence, heritability only accounts for, at best, half of the overall similarity. This is an important point that Judith Rich Harris draws our attention to in her book *No Two Alike*[39]. We are so used to thinking of identical twins as being identical, we fail to realize how different they actually can be. If you think about it, Daisy and Violet Hilton not only shared the same genes but they literally shared the same environment. How could they be so different?

Most people believe that one of the main reasons that individuals can be so different is because they were raised in different homes. The history of parenting is full of advice about how best to raise children and the bookstores have whole sections dedicated to parenting manuals. This comes from an understandable concern to look after our offspring and give them the best start in life as well as deep-seated beliefs about the forces that shape individual development. We have all grown up in a variety of households with different experiences that have shaped us, which is why there is a common assumption that we are what we are because of the way we were raised. When we blame delinquent children, we typically look to the parents. However, Harris spent many years surveying the fields of developmental psychology and concluded that when it comes to psychological outcomes such as intelligence and personality, neither genes nor the household environment can predict how we will turn out.

Ironically, that is a message that most parents probably do not want to hear, but they should be the first to agree with Harris. Any parent should be able to confirm that no matter how much they try to treat their various children equally, they end up very different. In fact, when the proper measurements are done, two siblings raised in the same household are not much more similar than two randomly selected individuals of roughly the same age plucked from the same population. Despite what most parents want to believe and parenting manuals promote, the home environment plays a relatively minor role in shaping the development of children.

If it isn't the home environment and it cannot all be the genes, then what explains individuality? Harris argues that the major determinant of a child's intellect and personality is the influence of their peer group – other children. While the child may behave according to their parents' expectations in the home, they put on a different face in the playground and shopping mall. Children act and respond to others differently in different situations. This is why children of immigrants do not learn their parents' accents when learning English, but adopt the local dialects and accents of the neighbourhood kids.

Harris's thesis is highly controversial as it goes against the modern trend for parenting expertise. It is also leaves out the extreme environments of Romanian orphanages and depressed mothers who have been shown to affect long-term development. Moreover, parents indirectly influence which peer groups children are exposed to because they choose the neighbourhoods and schools that their children end up in. That said, the goalposts are likely to shift again when one considers the pervasive role that social networking sites such as Facebook and Twitter now play in teenagers' lives. However, even if today's extensive networks outside the home play a greater role in shaping children, this cannot explain why Daisy and Violet, who shared the same genes, the same environment and the same peers, were still different. Perhaps it's because people treat identical twins, even those conjoined at the waist, differently so as to distinguish them. That seems plausible, but a more likely explanation is in itself unlikely – and that is the role of random events in development: an area of research known as *epigenetics*.

Epigenetics

What do the sex of a clownfish and the spread of the common cold have in common? A strange question maybe, but both are examples of epigenetic phenomena that are triggered by social behaviour. They both depend on the interaction of biology and the influence of others. Epigenetics is the study of the mechanisms of interaction between the environment and genes – the way that nature and nurture work together.

Epigenetics provides answers to the sorts of common questions we all ask ourselves. Are we born mad, bad or sad, or is our personality determined by events in our lives? Why are our children so different when we try to treat them equally? These questions are at the heart of how best to create the societies we wish to live in; often shaped and controlled by government policies and laws. The answers people prefer to give to these questions come from deep personal opinions and reflect their political persuasion about the role of the individual in society. However, epigenetics offers a new perspective to understand human development that combines our biology with our experiences.

As we noted earlier, genes are the strings of DNA molecules, found in every living cell, that instruct the cell what to become. They do this by building proteins from amino acids, which in turn are made from combinations of atoms of carbon, hydrogen, oxygen and nitrogen. Every cell in the body has thousands of proteins and DNA determines what type a cell is and how it operates by regulating the production of proteins. Genes are like books in a library that contain information that needs to be read or transcribed in order to build the proteins. The proteins instruct the cell to become something, such as hair follicles, while others can turn them into neurons. This is a very simplistic account and there is considerably more to the story of the mechanism of genes, but for the level of discussion here, it is sufficient to know that genes are like sequences of computer code within the cell that control its operation.

Genes build humans and humans are very complex animals. Each body is made up of trillions of cells and the initial speculation was that humans must have a considerable number of genes to code for all the different arrangements of cells in our bodies. In 1990, scientists working on the human genome project began to map the entire sequence of genes for our species, using sophisticated technology that enabled computers to read off the sequences as strings of code. Very soon, it appeared that initial estimates of over 100,000 genes had been way off. Although the project is still continuing, at the last count it would appear that humans have only 20,500 different genes. That may still sound like quite a few but when you consider that the humble fruit fly, *drosophila*, has 15,000 genes, humans look decidedly puny in the genetic endowment department. In fact, much simpler living things like the banana or the rather revolting roundworm have more genes than humans and, as if that were not enough, the organisms that have the highest and lowest numbers of genes are both sexually transmitted diseases, *trichomonas vaginalis* with 60,000 and *mycoplasma genitalium* with 517.

So the number of genes does not reflect the complexity of the animal. The reason we initially overestimated the number of genes for humans was because the role of epigenetics was not yet fully appreciated. Moreover, it turns out that there is more information encoded in the few genes we have than is ever actually used. Only 2 per cent of genes appear to be related to building proteins. This information is only activated when the gene becomes expressed and geneticists now understand that only a fraction of genes are expressed. In fact, gene expression is the exception and not the rule. The reason is that genes are sets of IF–THEN instructions that are activated by experiences. These experiences operate through a number of mechanisms, but genetic methylation is typically one that silences a gene and is believed to play a major role in long-term changes that shape our development. If you think about genes like books in a library and the library is the full genome, then each gene can be read to build proteins. Methylation acts a bit like moving a book out of reach so the information to build proteins cannot be read, or blocking access to it by placing some furniture in front of the book.

DNA may instruct cells how to form and organize themselves to build our bodies but these instructions unfold within environments that modulate their instructions. For example, the African butterfly *bicyclus anyana* comes in two different varieties, either colourful or drab, depending on whether the larvae hatch in the wet or the dry season. The genes do not know in advance, so are simply switched on by the environment.

Sometimes those switches are social in nature. For many fish, the social environment can play a fundamental role in shaping how genes operate, even to the extent of switching sex. Clownfish live in social groups that are headed up by the top female. What Pixar's film *Finding Nemo* did not tell the audience is that clownfish have the potential for transsexuality. When the dominant female in a school of clownfish dies, the most dominant male changes into a female and takes over. Or consider the humble grasshopper. When the population of grasshoppers becomes overcrowded, they change colour, increase in size and become gregarious and socially sensitive to other locusts. This transformation from a solitary grasshopper within a swarm is triggered simply by the amount of physical contact they have with others.[40]

Social environments can trigger a metamorphosis in a number of different species, but is there any evidence that social environments regulate human genes in a similar way? The example of the common cold helps to address the question. Social environments increase our susceptibility to the common cold but also influence how we fight it. Colds are more common in the winter months, not because of the lower temperatures (contrary to popular belief) but through the transmission of the virus between people. One reason why the virus may be more prevalent in the winter months is that we tend to congregate closer as the nights draw in, enabling the virus to transmit more readily from one to another. Viruses are small packets of DNA made up of about 10–100 genes that enter our cells and hijack the protein production to make copies of themselves. As this infection multiplies, the normal function of the cells and ultimately the whole of the body comes under attack. However, a virus's ability to express and duplicate its own DNA is regulated by our own body's reaction to social stress.

Social stress and isolation have long been known to affect viral infections, which is why we can all do with a little TLC along with our chicken soup when it comes to nursing a cold.[41] All this sounds like common sense, but what this folk wisdom reflects is an increasing understanding of the role of social factors in illness. An analysis of the DNA in the white blood cells or *leukocytes* of lonely adults revealed different levels of gene expression in comparison to adults who were not lonely.[42] Specifically, the genes responsible for producing antibodies to infection were downgraded, making their immune response less effective. This may explain why lonely adults are more vulnerable to diseases. What is remarkable is that the different gene expression is only found in those individuals who feel they are lonely and is not related to the actual number of social contacts they have. Even some of the most popular people can still be the loneliest in a crowd because it is how they feel that is more important, rather than the extent of their actual social circles.

If social factors can regulate the expression of viral genes, then our own complement of roughly 20,000 genes is likely to be regulated in biologically significant ways by social factors as well.[43] It is not only our biology but also our psychology that affects how we cope with illness.

Lamarck's daft idea

What is the evidence for epigenetic processes in humans? After all, humans do not spontaneously change sex when a dominant female leaves the group, but critical events can trigger changes in how our genes operate and sometimes the resulting changes in behaviour can be passed on to subsequent offspring. This is an astonishing idea but is not new. In the early nineteenth century a minor French noble, Jean-Baptiste Lamarck, proposed that characteristics acquired during a lifetime could be passed on to the next generation.

In support of this idea, he showed that the sons of blacksmiths had larger arm muscles than the sons of weavers before they ever took part in the family business, which he interpreted as an inherited characteristic. As another example, he suggested that giraffes' necks became long through their constant reaching up to high branches to eat leaves – a physical trait that they then passed on to their young.

Contrast this Lamarckian notion to Darwinian natural selection. In Darwin's theory there are two mechanisms that lead to change. The first is spontaneous mutation that generates variations among members of the group. Today, we now understand that this variation arises from genetic processes. Second, the environment operates to select those variations that endow the individual with a competitive advantage to breed and pass on the variation. With successive generations, the variant becomes stable in the population. In the case of giraffes, those born with a mutation that resulted in them having longer necks were more successful in breeding. It was not the experience of trying to reach leaves that was passed on to the offspring, but rather the genes that increased the length of the neck.

Darwin originally suggested that long necks would provide an advantage for reaching more leaves, but it turns out that there are a number of competing explanations.[44] What is known is that the mechanism of inheritance is not Lamarckian. Rather, long necks originated as a genetic mutation that was passed on while giraffes with short necks did not get the same opportunity to reproduce for some reason. Lamarckian theory has been roundly denounced as daft in scientific circles but epigenetics is casting new light on his ideas. Maybe experiences during a lifetime can influence the biology of the next generation.

There are so many problems and errors with Lamarck's evidence that it would be all too easy to consign the notion to the dung heap of bad ideas. Moreover, Darwin's theory of evolution by natural selection is simply better at explaining and predicting the data. And yet aspects of Lamarck's daft idea have been resurrected with the rise of epigenetics. Sometimes events during one's lifetime can affect the next generation. Epigenetics explains how environmental signals change the activity of genes without altering the underlying sequence of the DNA. The process of natural selection will ultimately correct any epigenetic influences of the environment. Rather, the effects are more to do with the switches that are being flipped by epigenetic processes. So Lamarck may have gained a minor battle, but Darwin has won the war in explaining how we pass on characteristics from one generation to the next. Epigenetics may even explain why humans traumatized as infants grow up with an emotional legacy that can stay with them for the rest of their lives. Once again, studies of the rearing practices of generations of laboratory rats have shown how early experiences shape the bond between mother and daughters.

Licking rats

What could be worse than licking a rat? For many people, rats are disgusting abhorrent pests associated with poverty, disease and death. This is rather unfair, as the female rat is an intelligent and social animal with a strong maternal instinct. When she is rearing her pups in the nest, the female rat will invest time licking and grooming her brood like an attentive mother. Some mother rats are much more conscientious, with very high rates of licking, whereas others are less so – a trait these mothers share with all their sisters.[45]

What is remarkable is that if you take female pups from a low-licking mother and have them raised in the litter of a high-licking mother, they will acquire this attentive trait. Likewise, if you cross-foster in the opposite direction, you get the opposite effect.[46] Is this rat example simply a case of learning how to raise your pups? There is more to it than that. Grooming and licking appears to regulate the baby rat's response to stress. Those mothers with a high licking rate produce offspring who cope much better with stress than those from a low-licking mother. They also grow up into more resilient adult rats and, if female, pass this behavioural trait on to the next generation.[47] They are better adapted to reproduce.

You can even generate this effect if rat pups are reared by humans and given different levels of handling during the early days. This activity changes the baby rats' HPA response by altering their reactivity to stress. The grooming and licking releases the 'feel-good' neurotransmitter serotonin that regulates the gene that controls for GR in the hippocampus. In contrast, this gene is switched off in the under-stimulated pups, whereas it is almost never methylated in the pups of high-licking grooming mothers. With higher levels of GR expressed in the hippocampus, the rat is better able to regulate the HPA effectively. Even though DNA methylation patterns tend to be stable, if you cross-foster the pups of high- and low-licking mothers during the critical period, you can reverse the methylation of the gene in the hippocampus. In short, the early grooming experience is turning the genes on or off.[48]

This may be all well and good for rats, but what of humans? Is there any evidence of biological embedding of early experiences later in life? Post-mortem examination of suicide victims revealed that GR expression in the hippocampus was reduced in those with a history of early abuse compared to those without this childhood trauma.[49] What makes this finding all the more incredible is that it was not the stress of events that ultimately led them to take their own lives that produced this genetic difference, but rather events during their childhood that were responsible for silencing the genes.

It should be noted that there is not just one gene responsible and there are a multitude of different types of stress that affect individuals differently. In a recent study[50] of teenagers whose parents had reported stress during their child's upbringing, methylation of environmental stressor genes was investigated. The effects of the mother's stress were only evident if that stress had occurred when the child was still an infant. Fathers also produced methylation in stress-related genes but only when the child was older, during the preschool years. More intriguing was the finding that this effect was restricted to the girls in the study. It has been reported for some time that absent or deadbeat fathers have a greater influence on their daughters than their sons, but this study is some of the first evidence to point the finger of suspicion at epigenetics.

Warrior genes

On 16 October 2006, Bradley Waldroup sat in his truck drinking heavily and reading the Bible. He was waiting for his estranged wife to arrive with their four kids for the weekend. When his wife, Penny, turned up with her friend Leslie Bradshaw, a fight broke out and Waldroup went berserk. He shot the friend eight times and then chased after his wife before hacking her to death with a machete. It was one of the worst, most bloody crime scenes that Tennessee police officers had ever dealt with. What makes this horrific crime stand out, apart from its sheer brutality, was that it was one of the first cases where defence lawyers successfully argued that Waldroup should not be given the ultimate punishment of the death sentence because of his genetic make-up. They argued that Bradley Waldroup had a biological disposition towards extreme violence because he carried a *warrior gene*.

This gene was discovered in the Netherlands in 1993 by a geneticist, Hans Brunner, who had been approached by a group of Dutch women who were concerned that the males in their family were prone to violent outbursts and criminal activity including arson, attempted rape and murder.[51] They wanted to know if there was some biological explanation. Brunner soon found that they all possessed a variant of the monoamine oxidase A, or *MAOA* gene located on the X chromosome. In the following years, evidence mounted to support the link between patterns of aggression and low-activity MAOA. The condition would have remained known as the rather lame 'Brunner Syndrome' if it had not been for the columnist Ann Gibbons, who christened it the 'warrior gene'.[52] The emotive name is a bit of a misnomer as it is more of a *lazy* gene because it fails to do its main job, which is to break down the activity of neurotransmitters.

With the discovery of the warrior gene, soon everyone was looking for this biological marker in the underclass of society. Male gang members were more likely to have the warrior gene and four times more likely to carry knives. In one particularly inflammatory report[53] based on a very small sample of males, New Zealand's indigenous Maoris, famous for their warrior past, were found to have the gene. Not surprisingly, this report created a public outcry.

One of the problems with these sorts of studies is that they relied on self-report questionnaires, which are often inaccurate. One ingenious study[54] tested the link between the warrior gene and aggressive behaviour by getting males to play an online game where they could dish out punishment in the form of administering chilli sauce to the other anonymous player. In fact, they were playing against a computer that was rigged to deliver a win to itself despite the best efforts of the male participants. Those with the low-activity MAOA gene wanted revenge and were significantly more likely to deal out punishments in retribution.

The warrior gene may be linked to aggressive behaviour, but as Ed Yong, the science writer, puts it, 'The MAOA gene can certainly influence our behaviour, but it is no puppet-master.'[55] One of the main difficulties in explaining violent behaviour like that of Bradley Waldroup is that around one in three individuals of European descent carry this gene, but the murder rate in this population is far less. Why don't the rest of us with the gene go on a bloody rampage?

The answer could come from epigenetics. Genes operate in environments. Individuals are more likely to develop antisocial problems if they carry the low-activity MAOA-L *and* were abused as children. Researchers studied over 440 New Zealand males with the low-activity MAOA gene and discovered that over eight out of ten males who had the deficit gene went on to develop antisocial behaviours, but only if they had been raised in an environment where they were maltreated as children.[56] Only two out of ten males with the same abnormality developed antisocial adult behaviour if they had been raised in an environment with little maltreatment. This explains why not all victims of maltreatment go on to victimize others. It is the environment that appears to play a crucial role in triggering whether these individuals become antisocial.

As for Bradley Waldroup, was it right for him to be treated more leniently? He certainly had an abusive childhood, but probably no greater than other males in that trailer park in a socially deprived area of Tennessee. A third of them presumably also carried the gene. At the time, Waldroup was drunk and we know that alcohol impairs our capacity to regulate rage and anger arising from the limbic system. But was he accountable for his actions?

What is clear is that it was the evidence for a warrior gene that persuaded members of the jury to find Waldroup not guilty of murder even though it is clear that they did not fully understand the nature of gene-environment interactions. After the Waldroup outcome, one might think that raising the presence of the warrior gene and abusive childhood would make for a good defence strategy in the law courts. However, there is another way of seeing the argument. One could argue that those with a genetic predisposition towards violent crime should not be let off more lightly but rather punished more severely. This is because they are more likely to reoffend and so the deterrent should be even harsher. Warrior genes and abusive childhoods are risk factors that make some more inclined to violence, but then again, punishment and retribution are also factors that reduce the likelihood of committing crimes. The decisions we make throughout our life represent the interaction of biology, environment and random events. Deciding which is more important is the job of society operating through its laws and policies, but it would be wrong to think that the answer is simple.

The landscape of life

It is a cliché but we often talk about life as a journey with many forks and turnings. Just think about where you are today and how you got here. Did you know ten years ago where you would be today? Although some things in life are certain (death and taxes) and some are likely, many events are unpredictable. And some leave a lasting impression.

As we develop from a simple ball of cells into an animal made up of trillions of cells, the process is guided by instructions shaped by natural selection over the course of evolution and encoded in our genes. However, the genome is not a blueprint for the final body but rather a script that can vary depending on events that take place during development. These are not only events outside the womb, but also those within the unfolding sequence of body building, which explains why identical twins end up different despite sharing the same genes. They may start out with the same genome, but random events can set them off on different pathways as the body is being constructed. This explains why there are increasing differences observed in genetic methylation for identical twins the older they become. Even cloned fruit flies raised in the same lab-controlled environment should be absolutely identical and yet have different arrangements in their brains. When you consider the complexity of neural networks and the explosion of synapses with googols of connections, it seems obvious that no two brains could ever be the same.

Despite the diversity in brains that is inevitable during development, evolution still produces offspring that resemble their parents more than they resemble another species. The majority of genetic information must be conserved and yet remain flexible enough to allow for individual variation arising from events that occur during the developmental process. One way to consider the influence of genes and environment is to think about the journey through life as an epigenetic landscape.

In 1940, the brilliant British polymath Conrad Waddington used a metaphor of a ball rolling down a corralled landscape made of up troughs and valleys of different depth. The diagrams overleaf represent two paths of development in two individuals (A and B) who have the same starting genotype, as in the case of identical twins. These two individuals therefore inherit the same probability of developing a certain phenotype – the expression of those genes into characteristics that emerge over one's lifetime. However, they may have different actual phenotypic end points, determined by chance events and environmental effects, especially at critical points. At each junction, development can take a different path but whether it stays on course depends on the depth of the gully. Some gullies or canals are very deep, so the ball has no option but to follow that trajectory and it would take a mighty upheaval to set it on another course. These are the genetic pathways that produce very little variation in the species. Other canals are shallower, so that the path of the ball

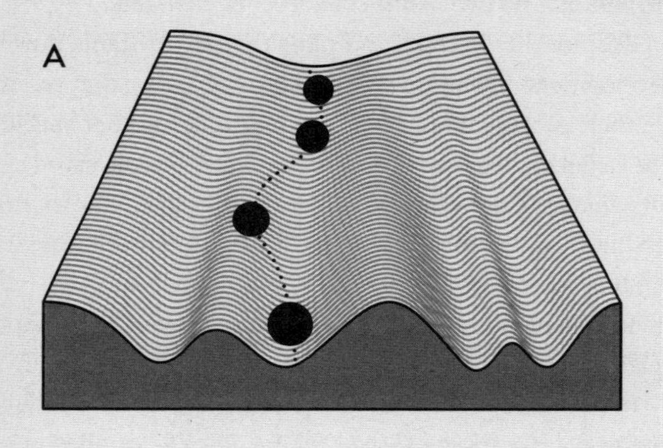

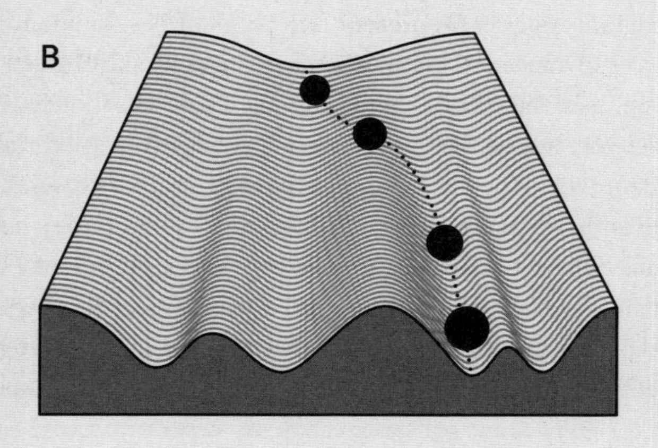

Figure 7: Waddington's epigenetic landscape
(After Kevin Mitchell, PL.S Bristol © 2007)

could be more easily set on another route by a slight perturbation. These are aspects of development that may have a genetic component but outcomes can be easily shifted by environmental events.

Waddington's metaphor of canalization helps us to think about development as a probabilistic rather than deterministic process. Most of us end up with two arms and two legs but it is not inevitable. Something dramatic during foetal development could produce a child with missing limbs, as happened during the 1960s when the drug thalidomide was given to pregnant women to prevent morning sickness. Other individual differences are much more susceptible to the random events in life that can set us off on a different course. This can happen at every level, from a chance encounter with a virus as a child to growing up in an abusive household.

Unravelling the complexity of human development is a daunting task and it is unlikely that scientists will ever be able to do so for even one individual, because the interactions of biology and environment are likelihoods and not certainties. There are just too many ways that the cards could stack up. More importantly, as the vernacular saying goes, 'Shit happens', which is a very succinct and scientifically accurate way of saying that random events during development can change the course of who we become in unpredictable ways.

CHAPTER 4

Who's In Control?

One day a scorpion and a frog met on a riverbank. The scorpion asked the frog to carry him across the water on his back because he could not swim. 'Hold on,' said the frog, 'How do I know that you won't sting me?' 'Are you insane?' the scorpion replied. 'If I sting you, then we will both die.' The frog, reassured by this, agreed to carry the scorpion on his back across the river. Halfway across, the frog felt the sharp, fatal puncture from the scorpion's tail. 'Why did you do that? Now we are both doomed,' cried the frog, overcome by venom, as they both began to sink below the water. To which the scorpion replied, 'I can't help it. It's in my nature.'

The story of the scorpion and the frog has been retold for thousands of years. It is a tale about urges and impulses that hijack our behaviours despite what is often in our best interests. We like to believe that we are in control and yet our biology sometimes gets in the way of what is best for us. As animals with the capacity to reason, we think that we are capable of making informed decisions as we navigate the complexity of life. However, many of the decisions we make are actually controlled by processes that we are not always aware of or, if conscious, seem beyond our volition.

Impulses are the drives behind the four Fs that we evolved to keep us alive long enough to reproduce. However, impulses and drives are not always appropriate – especially when they conflict with the interests of others. After all, there is a time and a place for everything. The impulses and drives that served us so well in our early evolution became problems in the modern era as we became domesticated. Being domesticated includes having to conform to social rules of what is acceptable in polite company. We may be highly evolved animals with more flexible behaviours than scorpions, but we also retain automatic urges that, unless regulated, can lead to self-defeat.

Many drives can become self-defeating. Some of us eat too much despite the warnings that we are ruining our health, whereas others starve themselves to death. Fighting often gets us into trouble and fleeing is not always the best thing to do when one should stand one's ground. Making unwelcome advances or engaging in sexual acts in public is not acceptable in decent society. Addicts knowingly abuse legal and illegal substances that will put them in an early grave. Chronic gamblers can squander away their families' futures and still believe that they can turn their luck around. There are those who feel the need to scream obscenities in public when it is the last thing anyone wants to hear. When we cannot stop these urges, we are no longer in control of our actions.

We all have the potential to be slaves to our urges and impulses and every so often we fail to keep them under control. How many of us have lost our tempers behind the wheel of a car, said things that should have remained as inner thoughts or acted in ways that we never thought we could?

In the cold light of day, we often know what we should think, say or do, but sometimes our urges and impulses get the better of us in the heat of the moment.

Scorpions may be rigid and inflexible, but humans possess a much greater capability to control our urges because we have evolved brain circuitry that plays a critical role in regulating our thoughts and actions. These mechanisms of self-control, shaped and strengthened by domestication, are essential for regulating behaviours in social settings. Without this self-control, we are in danger of being ostracized from the group.

The executive suite in our brain

The capacity for self-control is supported by neural mechanisms that traverse through the frontal lobes. Throughout human evolution, the frontal lobes expanded as our brains grew bigger and account for one-third of the human cortical hemispheres.[1] Although the human frontal lobes are bigger than those of the great apes, they are not proportionally larger than would be expected for an ape brain of human size.[2] However, as we noted in the preface, it is not so much the overall size but rather the way the microcircuitry of our frontal lobes is organized that is crucial to processing power. If you put a brain through a meat slicer, the cross-section reveals that there is more surface area where the cortical neurons are, tucked in the deep grooves and folds of the human brain in comparison to other apes. If you unfold the brain, humans would have more grey-matter surface area and therefore more potential for connectivity in the frontal lobes.[3] However, it is the change in this connectivity of the grey matter with experience that makes us so different from our closest primate cousins.[4]

In the adult brain, the frontal lobes are massively interconnected to most other regions of the brain and their size is largely due to the communicating neural fibres that make up the white matter beneath the grey-matter layer. However, this connectivity increases as the frontal lobes wire up to the rest of the brain. In comparison to other structures, the frontal lobes have the longest period of development, and during childhood they expand nearly twice as much as some other regions.[5] In comparison to apes and monkeys, who peak earlier in the synaptic explosion of growth that we described in Chapter 2, genes that control synaptic formation delay peak connectivity until as late as five years in humans, which explains why this brain region is the last to start wiring up.[6]

This delayed peak of activity in the wiring programme may have significant relevance to changes in behaviour. There is a noticeable shift in frontal brain activity between three and four years of age, when there is also a vast improvement in toddlers' abilities to plan and control thoughts and behaviours.[7] They become less impulsive, which may be a consequence of this changing brain connectivity which helps to regulate behaviours.[8]

Another fascinating role for the frontal lobes is that they enabled humans uniquely to imagine different possible futures – to mentally time travel and make plans for the future.[9] In fact, humans may be the only species that are able to contemplate the future.[10] Many animals, including rodents like hamsters and squirrels, can store and hoard food for the future, but these could simply be automatic reflexive behaviours that are triggered without much thought. Bonobos will carry around the correct tool for retrieving food for as long as fourteen hours, showing that they can anticipate the future for at least half a day.[11] But that's hardly the same as planning for next year's harvest. In one observation, cebus monkeys that were regularly fed once a day gobbled down as much food as possible until they were no longer hungry. They were always given more than they could eat in one sitting, but once they had eaten their fill, instead of saving it for another day they behaved like frat boys in a food fight and threw it out of the cage.[12]

Humans on the other hand plan for all sorts of future events. Much of our daily lives are taken up in preparation for anticipated outcomes. Our routines of schooling and employment are activities that pay dividends many years down the road. We even plan for our retirement decades in advance. Unlike most other animals, we can save for a rainy day. That level of foresight requires the integrity of the frontal lobes, which explains why only about a third of three-year-olds can tell you about what they are going to do tomorrow whereas twice as many four-year-olds can.[13] Immaturity and damage to these regions condemn us to living in the here and now, with little concern for how things might turn out.

The silent manager

The frontal lobes occupy an exalted position in the history of neuroscience.[14] The eighteenth-century Swedish scientist Emanuel Swedenborg first proposed that they were the seat of human intellect, a proposition that was later supported by the phrenologist Franz Gall in the nineteenth century. However, the activity of the frontal lobes remained surprisingly elusive to investigation. When Canadian neurosurgeon Wilder Penfield, who pioneered brain surgery on fully conscious patients in the 1940s, applied electrical stimulation to the surface of the brain, he noted how different areas triggered specific sensations or body twitches. In contrast, stimulation of the frontal lobes remained 'silent'. So what do the frontal lobes do?

What *don't* the frontal lobes do is probably the more useful question. Rather than being the focus of function, the frontal lobes are like New York's Grand Central Station, London's Heathrow airport or any large communication hub, receiving and sending information, linking up all the regions from the sensory systems, motor systems, emotional and memory regions spread throughout the brain. The massive amount of interconnectivity with the other brain areas indicates that the frontal lobes play a role in just about every aspect of human thought and behaviour. Rather than being localized in the frontal lobes, complex activities are integrated throughout this region like a neural junction box.[15]

Behaviour that requires planning, coordination and control enlists the activity of the frontal lobes. Even those activities that are automatic, such as the urges and impulses controlled deep inside the midbrain, need to be integrated into the rest of our behavioural repertoire so that they don't get us into trouble. One way to think about the frontal lobes is to imagine their role as like a senior executive management team overseeing a large company.

To be successful, a company must operate economically without wasting too much time and resources. The company needs to be able to take stock of the market, estimate demands, monitor current resources and set into action a planned strategy. The company will need to anticipate economic changes and plan for the future. Although there may be subdivisions in the company that compete for more resources than others, they have to be regulated so that the company as a whole can be more successful. This is why we need executives to manage the various operations that make the business run more efficiently as well as competitively. These executive functions monitor, coordinate, regulate and plan our thoughts and actions. Planning, memory, inhibition and attention are four executive functions (EFs) that operate from within a region that sits back from the front part of the brain, known as the *prefrontal cortex* (PFC).

Hot and cold

One useful distinction that has been made when considering the role of the PFC is the difference between 'hot' and 'cool' EFs.[16] Hot EFs include those impulses and urges that are biological imperatives or emotionally charged drives that threaten to take over control of our thoughts and actions, whereas cool EF refers to the logical choices that one has to make when presented with a problem to solve that requires rationality. We use cool EFs when we have to remember a telephone number or a list of things to buy from the store. Most of us will repeat the information over and over to keep it fresh in our minds before we forget. If the list of items is too long, we forget the beginning before we get to the end. The task is even harder if we have to remember two numbers or, worse still, if someone starts talking to us when we are trying to concentrate. Cool EFs enable us to keep focused on the problem. In contrast, hot EFs interrupt ongoing events and make us switch priorities. When the danger signs are detected, the hot EFs swing into action to protect us.

Developmental neuroscientist Yuko Munakata proposes that the PFC operates in two ways to regulate hot and cold decisions.[17] First there is the direct suppression of those drives and impulses by pathways that block the activity of mechanisms that are associated with hot, emotional EFs. Other thoughts and behaviours that represent the cold EFs that make up the normal routines of a typical day are regulated by indirect inhibition. These also need to be coordinated but without the need to shut down behaviours in the same way that hot EFs require. Munakata argues that this control is achieved by temporarily boosting the activity of different cortical regions. These support all the different options one is presented with when faced with a decision. In this competition, options with the strongest activation win out over those that are less active and so a decision is reached by the relative strength of different choices. Inhibition is not targeted at one behaviour in particular, but arises as a collateral effect of raising the profile of some options over others.

Changing your mind

Imagine that you are on an Easter-egg hunt out in the garden, where your goal is to find all the delicious chocolate-egg-hiding locations. So you set off and check under a bush here, or a tree there. But what if you could not remember where you had searched? You would end up returning to locations that you had already checked.

In our laboratory, we investigate children's searches using an automatic version of an Easter-egg hunt, where the goal is to check at all the locations that are lit up by pressing each light to see if it changes colour. A computer keeps track of all the searching, and we found that, below six years, children are very haphazard in how they go about the task and often return to locations they have already checked. They run around like headless chickens, drawn by each light even though they may have already checked it. The silent manager in their PFC is failing to coordinate and keep track of their behaviour. Rarely do they follow a systematic strategy.[18]

Easter-egg hunts might be a popular modern game but they are not too dissimilar to foraging. Not only did our ancestors hunt on the plains of the African savannahs, but they also foraged for nuts and berries. Remember how foraging requires a larger brain in the South American spider monkey in comparison to its close cousin the howler monkey, who simply loafs around eating leaves? Some of the extra brain tissue of the spider monkey is related to the need to remember locations and not make the mistake of returning to previous places they have visited. Even hunting requires remembering where you have been and not always going back to previous locations.

One way to be efficient in searching is by stopping yourself from returning to pastures old. This requires inhibiting the temptation to go back. Such flexibility to avoid doing

Figure 8: Searching for targets in our 'foraging' room

something is an important role of the frontal lobes that can be conspicuous in its absence. Adults with frontal-lobe damage can easily sort cards with coloured shapes into piles corresponding to either shapes or colours. However, if you ask them to sort according to one dimension such as colour and then get them to change in the middle of the task to the other dimension of shapes, they find it difficult to switch to the new strategy. They get stuck or *perseverate* in the response that was correct previously. What is more striking is that they can often acknowledge that they need to switch to the new sorting rule but still cannot stop themselves. They lack the ability to change their ways.[19]

Again this is a pattern that you see in normal development. Young children often get stuck in routines and indeed often seem to like repetition. It may be the familiarity they enjoy or it may be that they lack the flexibility to process changing information. This lack of flexibility is again related to the immaturity of the PFC, which no only keeps thoughts in mind, but also prevents us doing things that are not longer appropriate. This is why babies will continue to reach directly for a desirable toy in a clear plastic box only to discover that they cannot grasp it when their tiny hands bash up against the edge of the box.[20] Even though the side of the clear box is open and they can reach around to retrieve the toy, the sight of the goal is so compelling that they keep reaching directly for it. If you cover the object so that it is out of sight, they then learn to stop reaching directly for it. Something about the sight of the toy compels them to act. It's like showing an addict the fix they so desperately seek; they cannot avoid the temptation.

Even well-adjusted adults are not immune from doing the first thing that pops into their heads. One simple way to demonstrate this problem of inhibitory control is the *Stroop test* – a very simple task where you have to give an answer as quickly as possible in a situation where there is competition or interference from another response.[21] The most familiar version of the task can be found in a number of 'brain-training' games where you have to name the colour of ink that a word is written in. The task is easy if the word is 'red' and it is written in red ink. More difficult is when the word is 'green' but is written in red ink, because there is a conflict between naming the ink and the tendency to automatically read the word. Here is another Stroop task that you may not have encountered. Try counting the number of digits in each line as fast as you possibly can.

How many numbers are there in each row?

If you were answering as fast as you can, then you will have found the first four rows very easy but the next four much harder. You probably made mistakes, and if not, then you were probably much slower. Like the sight of a desirable toy for a baby, digits trigger the impulse to read them. As the digit conflicted with the number of items in some of the lines, the word had to be inhibited in order to give the correct answer.

If you break down a complex task into a list of things to do, then you can readily see why inhibition is so critical to performance. Some tasks need to be carried out in sequence, which is why I was hopeless at building model airplanes. I

5 5 5 5 5

3 3 3

2 2

1

3 3

5 5 5 5

4 4 4

2 2 2 2 2

am too impulsive for the task – something that was evident as a child when I always wanted to start painting my models before they were fully assembled. I lacked the patience that such hobbies require. This is why inhibition is necessary for planning and controlling behaviour by avoiding thoughts and actions that get in the way of achieving your goals. We can all experience this to some extent and it changes as we age. As we grow elderly, we become stuck in our ways. We lose flexibility of thought and can become more impulsive. Both the stubbornness and impulsivity are linked to the diminishing activity of our frontal lobes, which is just part of the normal ageing process. As we age, the PFC and its connections that regulate behaviours deteriorate faster than other brain areas.[22] At the beginning and the end of life, we lack the flexibility provided by our frontal lobes.

Disinhibited party animals

The EFs are not only important for reasoning, but play a vital role in our domestication that enables us to coordinate personal thoughts and behaviour with the wishes of others. They become facets of our personality that reflect the way we behave. Given their central role in shaping how we behave, you might imagine that the slightest damage to the PFC would immediately alter our personalities, but impairments in adult patients with frontal lobe damage are not always that easy to spot. Their language is usually intact and they also score within the normal range on IQ measures. However, frontal-lobe damage does change people in profound ways. Patients can be left unmotivated, with a dull, flat affect.

Others can become very antisocial, doing things that the rest of us find unacceptable because they no longer care about the consequences of their behaviour.[23]

Frontal-lobe damage may mean living for the moment, but this state is not as appealing as it might seem. Imagine if you suddenly no longer thought about getting on in life or how other people viewed your behaviour. Forget planning for the future and avoiding things that might get you into trouble. You would become reckless with money and people. You would do whatever took your fancy, no matter what the consequences. It would be hard to trust such a person. Frontal-lobe-damaged patients may seem normal, but they are often irresponsible, lack appropriate emotional displays and have little regard for the future. They find it difficult to tolerate frustration and react impulsively to minor irritations that the rest of us would let pass.

The most famous case of a change in personality following frontal-lobe damage is Phineas Gage, a twenty-five-year-old foreman who worked for the Rutland & Burlington Railway Company. On 13 September 1848, he was blasting away rocks to clear the way for the rails. To do this, a hole was drilled into the rock, packed with gunpowder, covered with sand and then tamped down with an iron rod to seal the charge. On that fateful day, apparently Phineas was distracted momentarily. He dropped the rod directly on to the gunpowder, igniting it to create an explosion that shot the 6ft metal rod through his left cheekbone under the eye and out the top of his skull to land 60ft away, taking a large part of his frontal lobes with it.

Remarkably, Phineas survived but he was noticeably changed in personality. According to the physician that looked after him, before the accident Phineas was 'strong and active, possessed of considerable energy of character, a great favorite with his men' and 'the most efficient and capable foreman'. After the accident, the doctor produced a report to summarize why the railway company would not re-employ him. Phineas was described as 'fitful, irreverent, grossly profane and showed but little deference for his fellows'. He was 'impatient of restraint or advice that conflicted with his desires'. In short, he had become a grumpier, ruder, more argumentative person, such that his former friends and acquaintances said he was 'no longer Gage'.

Due to the power of brain plasticity, Gage did eventually recover well enough to hold down another job as a stage-coach driver, but it is not clear whether his personality ever returned to that of the likeable fellow he had been before the accident. There has been considerable debate about Phineas Gage and whether his personality was permanently changed because the records at the time were poorly kept.[24] The story has been retold many times and something of a myth has been built up around this famous case. We have a much clearer picture with Alexander Laing, a former trooper in the British Army Air Corp, who is a modern-day Phineas Gage.[25] Following a skiing accident in 2000, Alexander suffered frontal-lobe brain damage that left him paralysed and unable to speak. He recovered quickly but on returning home became very antisocial, aggressive and unable to suppress his sexual urges. He is reported to have walked around his parents' house naked and acted inappropriately to women in public. At the time, his stepmother said, 'The damage to Alexander's frontal lobes seems to have exaggerated his character, although experts aren't sure if this is the case. I think the impulses were always there, but the lack of inhibition means he cannot control himself.' Of the time around his injury Alexander recalled ten years later, 'The frontal lobe damage was the worst. It meant I lost my inhibitions and did stupid things. It was like being permanently drunk. Afterwards I got into trouble of all sorts, I was even arrested twice. It was not a good time.'

Today, Alexander runs marathons for charity and seems to have got the better of his impulses, though his personality will probably never be the same as before the injury. During the 2011 London Marathon, he stopped after running 23 miles and began an impromptu dance in response to a Gospel choir performing at the side of the road to encourage runners, much to the delight of the gathered crowds. Only after the intervention of a medic assisting at the marathon was Alexander persuaded to stop dancing and return to the race. He believes that religion has kept him on the straight and narrow, showing that with the right social support, patients with frontal-lobe damage can experience considerable recovery. It also helps that we now have a better medical understanding of the importance of the frontal lobes in controlling our impulses. These cases reveal what happens to adults' social behaviours following damage to the frontal lobes. Understanding the relationship between EF and the frontal lobes helps explain why young children often behave in a way in which they seem oblivious to others around them and the embarrassment they create for their parents. Their immature frontal lobes have not yet been tuned up by the processes of domestication in the ways of how to conduct oneself in public.

Temper tantrums

'Daddy, I want it and I want it now!'

Who can forget Veruca Salt, the spoiled brat in Roald Dahl's *Charlie and the Chocolate Factory* who got everything she wanted? She may have been obnoxious but she really was no different from many young children when they cannot get their way. Around the end of infancy, children enter a phase that parents refer to as the 'terrible twos'. At this age, children have sufficient communication skills to let others know what they want but they are unwilling to accept no for an answer. It is a very frustrating time for parents because unless they give in to their child's demands the child can throw a temper tantrum – often for the benefit of a full public audience in a shopping mall or theatre. There is no point trying to reason with most two-year-olds because they do not understand why it is in their interests not to have what they want immediately. That requires the silent manager of the PFC to speak up.

One way to think about the PFC is that rather than supporting just one type of skill, it is engaged in all aspects of human behaviours and thoughts. As we grow older, our behaviours, our thoughts and our interests change. Situations that require some level of coordination and integration will require the activity of the frontal lobe EFs which do not reach mature levels of functioning until late adolescence.[26]

When adults have to learn a new set of information that conflicts with what they already believe to be true, there is heightened PFC activation during the transition phase, as revealed by functional brain imaging. One interpretation is that they are simply concentrating more, enlisting greater EF activity, but that activation depends on whether they have to contradict their initial beliefs. In this situation, the PFC activity is interpreted as reconciling incompatible ideas by inhibiting and suppressing knowledge that they previously held.[27] So rather than regarding any limited ability as being due to immaturity of the PFC, it is probably more accurate to say that changing behaviours and thoughts have not yet been fully integrated into the individual's repertoire – they are still learning to become like others.

In many social situations, young children think mostly about themselves, which explains why their interactions can be very one-sided. Some of us never grow out of this type of behaviour. These are the selfish individuals we all have encountered who only think about themselves. They do not care about what others think and behave as if their needs and opinions are the only important things in the world. They lack the patience and understanding that is required to have a balanced social relationship.

What happens to children who lack self-control when they grow up into adults? Terrie Moffitt and her team followed up over 1,000 children who had been born in the New Zealand city of Dunedin in 1972–3 and studied them from birth to the age of thirty-two years.[28] Each child was assessed for measures of self-control from three years of age based on reports from the parents, teachers, researchers and the children themselves. The results were startling. Children with high self-control were healthier, happier, wealthier and less likely to commit crime. These effects still held when intelligence and social background were taken into consideration. However, this was an observational study so it is difficult to know exactly what aspect of self-control was responsible for the outcome. What aspects of EFs were playing a major role in the children's entry into society? To answer that, we need a marshmallow, or possibly two.

Tempting marshmallows

In Germany, there was a tradition dating back to the Middle Ages that to determine whether a child was suitable for schooling you would offer them the choice of an apple or a coin.[29] If the child chose the apple, they had to remain in maternal custody. If the child chose the coin instead, they were considered 'worthy of instruction in knightly arts'. The logic behind the test was that a child who chose the apple was simply attracted by the desire to eat the apple, whereas a child who chose the coin could ignore the immediate gratification of the apple in favour of the greater rewards a coin would bring later.

This capacity to delay gratification has become a famous measure of children's self-control in a task known as the *marshmallow test*.[30] There are different variations of the test but they all involve presenting the child with a tempting reward. In the marshmallow version, children are instructed that the experimenter has to leave the room and that they can eat the treat now but if they wait until the experimenter returns they can have two – a much better deal but one that requires them to delay the gratification. This test has even entered popular culture as a UK confectionery manufacturer in 2011 ran an ad campaign using the same principle to demonstrate how tempting their products were.

Stanford psychologist Walter Mischel used the marshmallow test in the 1960s to measure how long children could delay gratification before giving into temptation. Around 75 per cent of four-year-olds failed to wait, with an average delay of six minutes. The most impulsive children simply gobbled the marshmallow up, but the others with more self-control resisted the urge. Not only did this measure indicate their capacity for self-control, but it predicted how well they got on with classmates and how well they performed academically when they were teenagers. It even predicted which males would later develop drug problems.[31]

All of these achievements of studying, getting on with others and avoiding drugs are components of domestication that require self-control. Studying can be boring and it is all too easy to find other, more interesting things to do. Getting on with others means that you have to be less selfish and more willing to share your time and resources. Avoiding drug use, a complex behaviour, requires at its heart the ability to simply say 'no'.

The marshmallow test seems to tap into an individual's natural impulsiveness. You might think that this self-control is all to do with the brain mechanisms for resisting temptation, but families also play a role. Different parenting strategies have been found to be associated with different self-regulatory behaviours in toddlers. In 1963, the famous child psychologist Erik Erikson wrote that 'the gradual and well-guided experience of the autonomy of free choice' will contribute to enhancing self-control, whereas over-control by the parent will produce the opposite effect.[32] In the decades following that statement, toddler research has generally provided support for this view. When asked to tidy up their toys, toddlers are more likely to be defiant if their mothers apply an over-control strategy of anger, criticism and physical punishment to control the child.[33]

One explanation is that children with strict parents have less self-control because they rarely have the opportunity to exercise their own regulatory behaviour, which is necessary in order to learn to internalize as a coping response. The process of domestication does not simply mean learning what the rules are, but when and how to apply them. This fits with classic studies that show that threats of punishment may work in the short term but persuasion is more effective when the potential threat is no longer around. Likewise, children of parents who use assertive persuasion exhibit more self-control because they have to develop self-regulation or suffer the consequences of being too impulsive. So the finding that less discipline leads to better adjustment runs completely contrary to the old saying that by 'sparing the rod, you spoil the child'. However, children of indulgent parents like Veruka Salt's also fail to exhibit self-control, indicating that allowing the child to run amok is also not a good strategy.

Of course, domestication strategies depend on how many children you are trying to raise. Clearly there is often competition between siblings and so single-child environments will be different from situations where there is more than one child. Whether such singletons differ from other children raised with siblings is controversial, but in China, which introduced a one-child policy in 1979, grandparents, teachers and employers believe that singletons are spoiled, selfish and lazy because they have been over-indulged by their parents. A 2013 study of these *Little Emperors* published in the prestigious journal *Science* found that children born just after the policy was introduced grew up into more selfish adults compared to those born just before.[34] They were also less trusting and helpful towards others.

The issue of trust plays an important role in our decisions to delay immediate rewards. After all, we are basing our decision on a promise that we will get something in the future. But what if a child is raised in an unpredictable environment where there is poor supervision as well as others who might steal their possessions or food? For these children, why should they take the risk if all they have known are broken promises? In this situation it would be foolish to wait. This reinterpretation of the delay of gratification finding is supported by studies on trust. If an experimenter promises but fails to give a sticker to four-year-olds in a drawing task, then these children are less likely to delay gratification on a subsequent marshmallow task.[35] It's not just children who do this. Adults will also forego an immediate financial reward for the promise of a greater reward in the future if they are told that the other person is trustworthy or indeed nothing is said about them at all. We are more likely to trust someone if we know nothing about them. However, as soon as the promiser is described as untrustworthy, then adults do not delay and take whatever is on offer. Our decision-making is greatly influenced by who we think we are dealing with.

These findings with adults seem obvious but they do cast a new light on understanding the relationship between self-control and sociability in at-risk children. Psychologist Laura Michaelson proposes that the classic relationship between failure of self-control as a child and later delinquency and criminality as an adult may actually reflect the lack of trust experienced at an early age as much as the biological factors that enable us to delay gratification.[36] Children from broken homes who grow up in impoverished households do not trust as much as those raised in supportive environments. No wonder they will take what they can get because, for them, the metaphorical bird in the hand is worth more than two in a bush.

If domestication means encoding our early experiences as contingencies in the neural circuits of the PFC, at the very least our capacity to learn from experience and moderate our behaviour suggests that self-control and life events probably work together in shaping our capacity for trust. Trusted adults strengthen the child's capacity for self-control. If a child has been told that they are 'patient' before the marshmallow test, they will wait significantly longer than children not given this label.[37] It may be a simple case of not giving a dog a bad name, as the old saying goes, but rather a good one. It is a simple *nudge* by an authority figure to help the child to strengthen its own resolve.[38]

Another revealing facet of self-control is what children do to regulate their behaviour. While waiting for the marshmallow, children who performed best on the task did not necessarily show more self-control but rather they seemed to find ways of taking their minds off the temptation. Many of them used distractions such as not looking at the marshmallow or singing to themselves. They were adopting a strategy known as *self-binding* – an action that one takes now in order to secure a better future. According to the Greek story, Ulysses wanted to hear the song of the Sirens but knew that their singing lured sailors to their deaths. To outwit them, he poured wax into the ears of his crew and had them bind him to the mast so that he would not leap from the ship and drown. Distraction turns out to be a better way of controlling urges because the act of resisting temptation by confronting it and trying to stop thoughts and behaviours can actually produce the opposite result in a psychological rebound effect.

Rebounding earworms and white bears

Rebound effects can happen when you least expect them and often can be very irritating.

Have you ever had that annoying experience where a tune gets stuck in your head – even one that you really hate? No matter how you try, it will simply not go away. The more you try to ignore it, the stronger the song becomes. Like some type of musical itch you cannot scratch.

This is because you are experiencing an *earworm*. Earworm is a direct translation of the German term *ohrwurm*, which means 'earwig'. These are the tunes that we can't forget, no matter how hard we try. It may be a catchy pop song or some advert jingle. Often we hate the tune but it simply will not go away. They intrude into our consciousness uninvited and, once there, overstay their welcome.

Around nine out of ten have experienced an earworm and diary studies indicate that most of us have an earworm episode at least once per week.[39] Most people find them annoying, but no matter how hard they try, these earworms just will not go away on command. And it is not just tunes that get stuck in our head; mental images can lodge in your mind as well.

You can assess your own mental-image suppression with the following test. Say out loud each thought or image that comes into your head over the next five minutes. Time yourself. You can say anything, but the only rule is that you must not think about a *white bear*. Remember that – anything but a *white bear*. Now try it.

Did the image of a polar bear pop into your mind? When my Harvard colleague Dan Wegner conducted this simple experiment, he found that participants could not help but think of a white bear and the more they tried to suppress the thought of a white bear, the more it rebounded back.[40] The reason for this obstinate effect is that in attempting not to think about the white bear, processes in our mind actively seek out white bears so as to monitor them and prevent them from entering awareness. However that monitoring in itself brings them into consciousness.

When people try to suppress unwanted thoughts, they come thundering back into consciousness with even greater strength. This failure of self-control can have implications for our domestication. Inappropriate sexual thoughts and racist stereotypes are both things that we would rather not think about, but in doing so, they become all the more vivid in our minds. In one study, adults were shown a picture of a skinhead and asked to write an essay about a day in the life of the individual portrayed in the photograph. Half of them were instructed not to use any stereotypes. After the essay, they were taken to a room with a row of eight empty chairs and told that the jacket on the end chair belonged to the skinhead they had just written about and were about to meet. Those who had suppressed the stereotype positioned themselves further away from where they thought the skinhead would be sitting than those who had not been given such instructions. This is the rebound effect in action. Even though these adults had not used stereotypes, actively suppressing the thoughts had altered their behaviour to make them even more susceptible to acting in a prejudiced way.[41]

Sometimes we cannot help ourselves, especially when our capacity for self-control has been compromised. After sustaining a concussion to his head, Basil Fawlty, the hapless hotel owner in the British classic comedy *Fawlty Towers,* was at pains not to mention the war when a group of German tourists came to stay. The more he tried to avoid mentioning the war, the more he let it slip during conversation. For children it may be marshmallows, but for adults it is all the thoughts and actions that we would rather not express in public because of the consequences they would have in terms of what others might think about us. Domestication means behaving in ways that are socially acceptable, something that requires sufficient self-control. Such self-control is difficult for young children, but for some adults, particularly those whose EFs are compromised by damage, disease or drugs, it continues to represent a considerable challenge.

Filth, harm, lust and Jesus

For some individuals, intrusive thoughts and behaviours completely undermine their ability to behave appropriately in social situations. Impulse control disorder (ICD) covers a variety of conditions acquired through disease and injury as well as those that arise during development. Phineas Gage and Alexander Laing had acquired ICD from frontal damage and there are various forms of dementia resulting from brain disease of the frontal lobes that produce syndromes where behaviour becomes inappropriate. However, for some individuals, they are born with ICD that impairs their social functioning.

One developmental disorder, named after French neurologist Georges Gilles de la Tourette, that has become synonymous with ICD is *Tourette Syndrome* (TS). TS is a condition characterized by involuntary thoughts and behaviours. These can be body jerks but they include vocal tics, from simple grunts to shouting obscenities in public or *corporallia*. This is often how they come to the attention of others, because strangers fail to understand that these individuals are unable to control their impulses. To someone who is not aware that an individual has TS, this can seem like the height of rudeness, which is why TS sufferers often end up in difficulty in social settings.

TS is a spectrum disorder that first appears around school age, increases during pre-adolescence but, for most, declines by the beginning of adulthood. The incidence may be as many as one in a hundred children, is more common in males than females and runs in families, indicating that it is a developmental brain disorder with a genetic basis. The typical symptoms relate to impulse control, which supports the idea that ICD must be related in some way to the PFC. This link has been confirmed by imaging studies that reveal that the connectivity of the PFC to an area of the brain that regulates behaviours known as the *basal ganglia* is altered in persons with TS.[42]

Those with TS fight a constant battle to inhibit their tics, especially in public, which usually makes the condition much worse, just like Basil Fawlty trying not to mention the war. As the pressure to behave normally in a social situation increases, the urge to tic increases, which makes it build up like a sneeze. And just like a sneeze, it becomes involuntary so that they must tic in order to get some relief. As one boy, Jasper, with TS explained on a HBO television special, 'When I try to hold back too much, you can't think of anything except holding them back and you can't think of anything except doing them.'[43]

Similar intrusive thoughts are also reported in individuals with *obsessive-compulsive disorder* (OCD), another ICD that affects around two out of every hundred adults in the West.[44] Obsessions are the tormenting thoughts whereas compulsions are the activities that the sufferer must engage in to counteract the obsession. If I am obsessed by thoughts of filth then I may feel the compulsion to wash my hands repeatedly.

Sir Aubrey Lewis, an English psychiatrist, described how obsessions generally fall into one of four categories: thoughts related to filth, thoughts related to harming oneself or another, thoughts about sex and the urge to blaspheme. What makes all these ICDs relevant to social acceptance is that all of these topics are associated with behaviours that need domestication. Inappropriate and excessive filth, harm, lusting and blasphemy would be frowned upon and so there is a need to keep such thoughts and behaviours in check. Currently the same inhibitory circuitry of the PFC and basal ganglia implicated in TS is also a prime suspect for OCD.[45] Like TS, there is a heritability factor, with OCD running higher in families, and it is more common in identical than non-identical twins. Not surprisingly, around half of those with TS exhibit obsessive-compulsive behaviours.

Many of us have intrusive thoughts from time to time or engage in peculiar habits. It could be our morning bathroom routines or it might be the coffee break that we always take at the same time of the day at the same coffee shop. Habits and routines are part of normal life, but we can happily switch or stop them should the need arise. They don't get in the way of us living our lives. However, individuals with ICDs can be ostracized from normal social integration. For many ICD sufferers, the worst aspect of their condition is not the disabling nature of their thoughts and behaviours, but the stigmatizing shame and embarrassment they can feel in public.

It wasn't me but the wine talking

Most of us lose control occasionally and, for many, letting go is part of being sociable. Otherwise we are uptight, too rigid and too inhibited. This is one reason why people drink alcohol. Contrary to popular misconceptions, alcohol is not a stimulant that turns someone into a party animal, but rather a depressant that weakens the inhibitory capacity of the frontal lobes, thereby unleashing the wilder, undomesticated animal with all its untethered drives. That is why we eat more, get into fights, lose reason and become more sexually active when we are drunk. After a night of misguided behaviour, many people wake up and explain that 'I was not myself' or 'It was the wine talking'. Of course, wine doesn't talk and if you were not yourself, then who were you?

As a domesticated species that has evolved to get on with each other, we must be careful not to insult or upset other members of our group. However, some of us harbour illicit thoughts and attitudes that are best kept to ourselves. If we care about what others think, then we try to keep all of these stereotypes, biases, drives and mistaken beliefs under wraps because we know that they are unacceptable. We may even understand that they are wrong but nevertheless they linger in our unconscious. However, just like white bears and earworms, the more we try to suppress these negative aspects of our self, the more they can rebound back despite our best efforts.

Whether we can suppress unwanted thoughts or not, it does call into question what our true nature is. Is it the inner secrets that we keep under lock and key in our mind, or the public persona we share with the world? Most of us probably would prefer to know about someone's secrets because ultimately we would be suspicious that individuals were not being honest about their true self. Even though someone might successfully hold back unpleasant aspects of their personality, the danger is that they might not always be able to control them.

The cost of control

Have you ever come out of a stressful exam or interview and thought that you could devour a whole tub of ice-cream? Or maybe it was an emotional movie that left you drained. Why do many people who endure a stressful experience want a stiff drink or to raid the fridge in search of comfort foods that are high in fat and sugar? One intriguing idea is that when we succumb to these temptations, we are experiencing ego depletion.

Ego depletion comes from American psychologist Roy Baumeister, who believes that enduring something stressful exhausts our capacity for willpower to the extent that we give in to temptations that we would rather avoid.[46] In one of his studies, he made hungry students eat bitter radishes rather than delicious chocolate cookies.[47] Even people who like a bit of radish in their salad would find that task difficult. However, Baumeister was not interested in eating habits. He was really interested in how long the students would persevere on an insoluble geometry task. The students who had been allowed to eat the cookies stuck at the geometry task on average for about twenty minutes, whereas those who were forced to eat the radishes gave up after only eight minutes. They had used up all their willpower to eat the radishes so they were left with less reserve to cope with another situation of completing a difficult problem.

Performance on one task that requires effort can therefore have unforeseen consequences for a subsequent situation that is completely unrelated except that it requires effort. This is why Baumeister regards willpower as a mental muscle that can become exhausted. We apparently spend quite a bit of time avoiding temptations. For one week, German adults carried around BlackBerries that quizzed them once in every two-hour period what they were thinking about. They were found to spend an average of three to four hours of each waking day avoiding temptations and desires.[48]

Just maintaining one's composure can be ego depleting. Not being allowed to laugh at hilarious comedy sketches, firing employees, enduring others in crowds are all situations where we have to exert self-control that leads to ego depletion. We exhibit more ego depletion at the end of the day, which is when couples are more likely to fight, after a hard day at the office. We become less tolerant of others and blame our spouses for the problems that are really generated by work.

When we are ego depleted, we eat more junk food, drink more alcohol, spend more time looking at scantily clad members of the opposite sex and generally have less control over our behaviour. Not only do we give into temptation, but we have an increased desire for forbidden fruit.

There's no one in control

Most of us believe that we are in control. We may dilly and dally about making decisions, but we still think that we are the ones making the choices. We feel the authorship of actions and ownership of thoughts. And yet we sometimes surprise ourselves when we do things that seem so out of character. It's as if we have an inner scorpion determined to behave the way it wants to.

We must keep these beasts at bay. In order to be domesticated, we must be able to control ourselves and learn when and where behaviour is appropriate. This self-control is our capacity to regulate and coordinate competing drives and urges. It develops over childhood, supported by the executive control mechanisms of the PFC that act to suppress and inhibit thoughts and actions that may potentially sabotage our goals. By observing others and learning what is appropriate, children learn to engage their self-control. When these control mechanisms are impaired, individuals are at the mercy of automatic thoughts and behaviours. Moreover, they are unable to foresee the consequences of their actions and become impulsive, trapped in the moment of immediate gratification.

If impulse control emerges as the interaction between biology and environment, it would seem that it is wise to provide children with guidelines about what is socially acceptable but not try to enforce them by external pressure. Nor should they be left alone or indulged. One size does not fit all and strategies for domesticating children will depend on the individual child, the parents and the culture. This variation in impulsivity reflects both individual temperaments but also the social environments that foster strategies for shaping and modifying thoughts and behaviours. If our social interactions are to be successful, we need to maintain control in company but that comes at a cost. When we resist the temptation to act inappropriately or not say what is on our mind because we might offend or upset others, then there can be consequences. Rebound effects and ego depletion show that there can be a price to pay for maintaining a veneer of respectability and when disease, damage or drugs compromise our self-control, we become victims of unconscious thoughts and behaviours as the story of the coherent individual that we try to maintain comes apart. When someone loses control, they often end up in trouble since they break moral codes and laws that society has put into place to guide our domestication. But what if there were no rules? Would we still learn to live together or would all hell break loose?

CHAPTER 6

The Longing

E

mo

time,

the rou

Shane Bauer was one of three American hikers imprisoned in Iran in 2009. At the time of their arrest in the Middle East, Shane, his girlfriend Sarah Shourd and friend Josh Fattawere were hiking in the Zagros Mountains of Iraqi Kurdistan, looking for the Ahmed Awa waterfall, a tourist attraction near the Iraq–Iran border. After they visited the waterfall, the Iranian authorities claimed that they had entered Iran illegally and arrested the three on suspicion of spying. Shourd was released after fourteen months on humanitarian grounds, but Bauer and Fattawere were convicted of espionage and sentenced to eight years' imprisonment. They spent twenty-six months in captivity and were later released in September 2011 after bail of $500,000 was paid.

This experience in a foreign land would leave a profound effect on Bauer and his attitude towards imprisonment, especially when he discovered that prison conditions were sometimes more extreme in his own country. In an article in the magazine *Mother Jones*[1] Bauer wrote, 'Solitary in Iran nearly broke me. I never thought I'd see worse in American prisons.' He was determined to reveal the horrors of his homeland's use of solitary confinement as a form of legalized torture. On a visit to a Californian prison, an officer asked him about his time in Iran. Bauer explained

> no part of my experience – not the uncertainty of when I would be free again, not the tortured screams of other prisoners – was worse than the four months I spent in solitary confinement. What would he say if I told him I needed human contact so badly that I woke every morning hoping to be interrogated?

Loneliness is often only a temporary state as one adapts to new environments, but when that isolation is used as a punishment enforced over days, months and even years in solitary confinement, it can be the cruellest way to treat another human. Physical torture and starvation are dreadful, but according to those who have suffered imprisonment, it was the isolation that they found the worst. Of his time in prison on Robben Island, Nelson Mandela wrote that 'Nothing is more dehumanizing than the absence of human companionship', and he knew men in prison who preferred half a dozen lashes with a whip rather than being in solitary confinement.[2]

It is estimated that 25,000 US prisoners are currently locked in tiny cells, deprived of all meaningful human contact. Many of them spend a few days there. Some have been isolated for years. These are not always the most violent inmates. Prisoners have been 'locked down' for simply reading the wrong book. There are no international codes of conduct for this punishment and no other democratic country uses solitary confinement as much as the US. It is a shocking anomaly from a nation that claims to be so committed to human rights. In 2012, the New York Civil Liberties Union published their findings about the use of solitary confinement in the state and concluded 'These conditions cause serious emotional and psychological harm, including severe depression and uncontrollable rage.'[3]

Those who willingly volunteer for isolation can also experience psychological distress. Forty years ago, French scientist Michel Siffre conducted a series of studies to investigate the rhythms of the body when isolated from external measures of time such as natural sunlight. He spent months in caves without any clocks or calendars and discovered that the human body operates not on a twenty-four-hour cycle, but rather on a forty-eight-hour cycle when there are no daylight cues. Given enough time in isolation, people will revert to a cycle where they stay awake for thirty-six hours and then sleep for twelve.[4] He also discovered the psychological pain of social isolation. Even though he was in constant communication with his assistants above ground, his mental health began to deteriorate. In his last study, conducted in a cave in Texas, he began to lose his sanity.[5] He became so lonely that he tried to capture a mouse that he had named *Mus* that occasionally rummaged through his supplies. Siffre wrote in his diary,

> My patience prevails. After much hesitation, Mus edges up to the jam. I admire his little shining eyes, his sleek coat. I slam down the dish. He is captured! At last I will have a companion in my solitude. My heart pounds with excitement. For the first time since entering the cave, I feel a surge of joy. Carefully I inch up the casserole. I hear small squeaks of distress. Mus lies on his side. The edge of the descending

> dish apparently caught him on the head. I stare at him with swelling grief. The whispers die away. He is still. Desolation overwhelms me.

Such is the need for companionship that the audience can fully understand why the shipwrecked FedEx employee Chuck Noland played by Tom Hanks in the movie *Cast Away* (2000) strikes up a relationship with a volleyball he calls Wilson (after the ball's manufacturer). He even risks his own life to save Wilson when the ball falls into the ocean during an attempt by Chuck to escape the island on a makeshift raft. Chuck dives into the ocean after the ball, calling out desperately for Wilson, but eventually gives up, apologizing to the ball as it drifts off on the current. It is one of the most unusual 'death' scenes for an inanimate object and yet this emotional trauma immediately resonates with the audience because we understand what loneliness can do to someone.

Just like me

These tales of desperation for companionship reinforce a central point of this book: that the human brain evolved for social interaction and that we have become dependent on domestication for survival. Social animals do not fare well in isolation and we are the one species that spends the longest period being raised and living in groups.[6] Our health deteriorates and life expectancy is shortened when we are on our own. The average person spends 80 per cent of their waking hours in the company of others and that social time is preferred to time spent alone.[7] Even those who deliberately seek out isolation, such as hermits, monks and some French scientists, are not exceptions that prove the rule.

It is not enough just to have people around; we need to belong. We need to make emotional connections in order to forge and maintain those social bonds that keep us together. We do things to make others like us and refrain from doing things that make them angry. This may seem trivially obvious, until you encounter those who have lost the capacity for appropriate emotional behaviour and you realize just how critical emotions are for enabling social interactions. Various brain disorders such as dementia can disrupt emotions, making them too extreme, too flat or too inappropriate. Even those without brain disorders vary in their capacity for emotional expression. Those lacking in or unwilling to share their emotions are cold and unapproachable, whereas others who willingly express their emotions, assuming they are positive, are warm and friendly.

Sometimes others' emotions can be infectious. Emotional contagion describes the way that others' expressions can trigger our emotions automatically. Many of us get teary when we see others crying at weddings and funerals. Or we may collapse into a fit of the giggles when a friend does, even though we should be keeping our composure in front of others. Actors call this 'corpsing', probably because the worst time to giggle is when playing a corpse on stage.

Laughter and tears are two social emotions that can transmit through a group like an involuntary spasm. When we are sharing these emotions we are having a common experience that makes us feel connected to each other. We know this is innate, rather than learned, because babies will also mimic the emotions of others. They cry when they hear other babies cry or see others in distress. Charles Darwin described how his infant son William was emotionally fooled by his nurse: 'When a few days over 6 months old, his nurse pretended to cry, and I saw that his face instantly assumed a melancholy expression, with the corners of his mouth strongly depressed.'[8]

What could possibly be the benefit of emotional contagion and why do we mimic some expressions and not others? One suggestion is that expressions evolved as adaptations to threat. Fear changes the shape of our face and raises our eyebrows, so that can make us more receptive to potential information from the world. On the other hand, disgust, where we wrinkle up our noses and close our eyes, produces the opposite profile, making us less susceptible to potentially noxious stimuli.[9] Seeing or hearing someone vomit makes us gag, possibly as a warning to expel the contents of our own stomach as we both may have eaten something that is not good for us.

Our capacity for imitation is supported by brain mechanisms that form part of the so-called *mirroring system* – a network of brain areas that include neurons in the motor cortex that control our movements. These neurons are normally active when we are planning and executing actions. However, back in the 1990s in Parma, Italy, researchers chanced upon a discovery about motor neurons that was to change the way we think about ourselves and what controls our actions. Vittorio Gallese and colleagues had been measuring from a neuron in the premotor cortex of a rhesus macaque monkey, using a very fine electrode.[10] The cell burst into activity when the monkey reached for a raisin. That was to be expected as it was a premotor neuron that initiates movements. However, the Italian researchers were astonished when the same cell also fired as the monkey watched the human researcher reach for a raisin. The monkey's brain was registering the experimenter's reaching; an activity that was controlled by the human brain.

The reason this is remarkable is that it used to be thought that the areas for perceiving others' actions were different from the network for producing your own movements. Instead, the Italian researchers had discovered that around one in ten neurons in this region were 'mirroring' the behaviour of others. It was as if these mirror neurons in the monkey's brain were pantomiming the actions of others. As neuroscientist Christian Keysers explained, 'Finding a premotor neuron that responds to the sight of actions was as surprising as discovering that your television, which you thought just displayed images, had doubled all those years as a video camera that recorded everything you did.'[11]

This dual role of copying other people's behaviour and executing your own set the scientific community alight. Direct mapping between our brain and the brains of others, by observing them, could explain why we cry at weddings, feel others' pain, emotional contagion and all manner of social behaviours that seem to reveal the human capacity for mimicking. It was as though scientists had found a direct psychic connection between the minds of others. It was even announced that the discovery of mirror neurons was as significant to understanding the brain as the discovery of the structure of DNA was to biology; while this is an exaggeration, it captures the excitement mirror neurons generated.[12]

Others were more sceptical because recording directly from neurons in the brain of a human had not been done. However, in 2010, neurosurgeon Itzhak Fried published a study[13] of patients he had been treating for epilepsy. To isolate the affected brain region, he implanted electrodes to determine which areas to surgically remove – much in the same way that Wilder Penfield had done all those years earlier with his neurosurgery patients. During this procedure, the patients were fully conscious and able to take part in a study designed to establish the presence of mirror neurons once and for all. They were asked to either smile, frown, pinch their index finger and thumb together or make a whole grip with their hand. When Fried found neurons that were activated during one of these movements, the patients were then shown a video of someone else making the same types of movements. Just as in the macaque monkey, premotor neurons were activated both by making a movement and also by watching someone else perform exactly the same action – *bona fide* mirror neurons in humans. The real burning question is how did they get there?[14] Are they simply neurons that have acquired their dual activity after years of watching others and mapping their behaviour to one's own movements? Or are babies already prepackaged with mirror neurons, which might explain reports where newborns have been shown to copy adult facial expressions without any learning?

The 'in' crowd

As we read in Chapter 2, there are reasons to believe that we may be born with a rudimentary capacity for mimicking others. Infant mimicry is instinctual but the system is not simply a dumb mechanism that slavishly copies every Tom, Dick or Harry a child encounters. Rather, infants become more discerning of others, assessing whether they are friend or foe. Initially, this distinction is drawn between those that share the same interests and preferences as the baby. In a food-preference study,[15] eleven-month-olds were offered the choice of crackers or cereal from two bowls. Having made their choice, they watched as two puppets came along and approached the food. For each bowl, one puppet said, 'Hmm, yum, I like this' and the other said, 'Ewww, yuck, I don't like that.' Each puppet expressed the opposite attitude to each food. The infant was then offered the choice to select to play with one of the puppets. Four out of five infants chose the puppet that had the same food preference as him or herself, irrespective of whether it was crackers or cereal. Before they have reached their first birthday, babies are showing clear signs of preference and prejudice. Just as their brains are tuning into the faces and voices that surround them, so too are they learning to identify who is, and who is not, like them.

To make this distinction, one has to have a sense of self-identity – knowing who we are and how we differ from others. This emerges most conspicuously during the second year of life. Famously, humans and other social animals recognize themselves in mirrors.[16] Initially young infants treat their reflected image as a playmate, but around eighteen to twenty months they begin to show reliable mirror identification, indicating a new level of self-awareness.[17] Somewhere between two and three years of age, children begin to show signs of embarrassment as indicated by blushing. As blood flushes our skin, reddening our face, blushing is an indicator of being uncomfortable in a situation that attracts the undesired attention of others. As Charles Darwin noted,

> It is not the simple act of reflecting on our own appearance, but the thinking what others think of us, which excites a blush. In absolute solitude the most sensitive person would be quite indifferent about his appearance.[18]

Why blushing evolved is a bit of a mystery, but one suggestion is that it could work as a visual apology to others, thereby averting social ostracism.[19] The problem with that is that blushing is not obvious in dark-skinned people and we were all dark once. Did it evolve its signalling properties only after the migration out of Africa? Nobody really knows why humans are the only animal that blushes, but the fact that it only occurs in the company of others means it must be related to signalling our sense of shame and guilt – emotions that depend on what we think others are thinking about us.

Self-awareness in children is also signalled by the appearance of the use of personal pronouns that we talked about in the last chapter when it comes to owning stuff. Towards the end of the second year, children are using 'I', 'me', 'mine', but they are also using gender labels such as 'girl', 'boy', 'man' and 'woman' – though females are ahead of males simply because they are generally more advanced with language.[20] This self-labelling as a boy or girl is one of the first markers of identity. Infants are sensitive to gender much earlier because they all show preferences for the female face at three to four months of age,[21] but by the time they are two years old, most have a preference for their own gender.[22] In fact, sensitivity to gender predates racial prejudice that appears much later. When asked to select potential friends from photographs, three- and four-year-olds show a reliable preference for their own gender but not their own race.[23]

Once they know they are a boy or a girl, they become *gender detectives*, seeking out information about what makes boys different from girls.[24] This is when they begin to conform to the cultural stereotypes present in society. Not only are they gender detectives but they also police the differences as enforcers, criticizing those who display attitudes or behaviours associated with the opposite gender. By three to five years of age, children are already saying negative things about other children who they do not identify with. They are making a distinction between in-groups and out-groups. If you are in my gang, then we are both in-group members.

Initially group identity is gender specific but it can be based on something as trivial as dress code, which is why three-year-olds will prefer other children who wear the same-coloured T-shirt as themselves.[25] Child psychologist Rebecca Bigler at the University of Texas in Austin, who has spent twenty-five years studying interventions aimed at countering children's bigotry, has concluded that once the child develops a prejudicial social stereotype, it can be almost impossible to get them to abandon it – 'In the case of stereotyping and prejudice, it may well be that an ounce of prevention is worth a pound of cure.'[26]

Knowing me, knowing you

When we think about ourselves and others, certain areas are activated in our brains. Harvard neuroscientist Jason Mitchell,[27] one of the new vanguard of researchers in the field of social cognitive neuroscience, has pointed out that there is good evidence that there may be four to six regions that form networks that are consistently activated in social situations and not in other types of problem solving. If you are asked to imagine whether or not a historical figure like Christopher Columbus would know what an MP3 player is, this type of question activates these socially sensitive networks. This is because you have to infer the mindset of Columbus and imagine what he would think. However, if you ask whether an MP3 player is smaller than a bread bin, these areas remain silent. This is because the question now becomes a perceptual judgement based on your knowledge about the size relationship between different physical objects.

One of the circuits activated by social encounters is the mirroring system and includes regions where the mirror neurons we mentioned earlier have been found. This circuitry registers the physical properties of others as well as our own body shape and movements. It includes the premotor areas, parts of the frontal cortex and the parietal lobes – all regions involved in actions. The integration of neural systems that represent both our own bodies and the bodies of others can explain why watching the suffering of another person when they are in pain also triggers our own corresponding brain regions.[28]

In addition to a system that registers the physical similarities with another person, another circuitry is activated when we contemplate ourselves, compared to thinking about others. This mentalizing system involves the medial prefrontal cortex (MPFC – the region in the middle of your forehead), the temporal-parietal junction (TPJ – where the two lobes meet, which is a couple of inches above the temples) and the posterior cingulate (PC – located close to the crown of the head) and appears to support thought processes when you are thinking about yourself. These contemplations include the relatively stable aspects of our personality that we have insight into, such as 'I am really quite an anxious person', as well as ever-changing feelings such as 'I am feeling quite confident at the moment'. This circuitry is also active when we mentally time travel to think about the past or imagine our self in the future.

Both stable and transitory self-reflection show up as increased activation of the MPFC.[29] Self-reflection also includes the extended self of objects. In the same way that there is a characteristic P300 brain signal that registers stuff that belongs to you, as discussed in the last chapter, the MPFC is activated in situations where the endowment effect is triggered, which supports the idea that this region is part of the neural representation of at least one aspect of self.[30]

The self-reflection system is not just about naval gazing, however. It enables us to imagine our self in different situations that others might face. This sort of ability would enable one to self-project or simulate another person's situation possibly as a way of understanding what their thought processes or emotions might be. The Scottish social neuroscientist Neil Macrae has described the MPFC system as a kind of *knowing me, knowing you* mechanism. In other words, when you are making judgements about other people, you are really comparing them to yourself. This is why, when adults are asked to judge others, the more objectively similar they are to the person they are thinking about, the more activation is observed in their own MPFC.

Once we identify with others in our group, we are more likely to mimic and copy them. These are acts of affiliation, signalling our allegiances. We want to be seen to be like others in the group in order to consolidate our position. However, if someone from an out-group copies us, we interpret this mimicry as mockery – an act of provocation. It is not enough that we like those who like us, but we are actively suspicious of others who are not in our tribe.[31]

Our empathy is also two-faced. When we watch someone from our ethnic group receive an injection in the cheekbone, we wince and register more mirrored pain in our brains compared to watching the same pain inflicted on someone from another race.[32] We can more easily watch others suffer if we do not identify with them. Taken to its logical conclusion, we can witness and inflict suffering upon others without feeling any remorse by dehumanizing them. This is one of the reasons why we refer to those we persecute as insects, parasites, animals, plagues, or any other term that demeans our enemy or victim as not being a member of the human race.

When the division between groups escalates into conflict, humans treat each other in the most terrible ways imaginable. Whether it is political, economic or religious justification, there seem to be no boundaries when it comes to the suffering and cruelty we can inflict upon other humans when we regard them as the enemy. This has been borne out in countless conflicts in the modern era, where neighbours have turned on each other and committed atrocities that seem inconceivable. Cambodia, Rwanda, Bosnia and Syria are just a few examples where communities that had known decades of peaceful coexistence suddenly erupted into genocide as one group tried to obliterate another.

That ordinary people can readily commit extraordinary atrocities against their neighbours is puzzling. What can make people behave in such a way that one would never dream possible? One explanation is that our own moral code is not as robust as we would wish. We are not as independently minded as we think we are. Rather, we are easily manipulated by the influence of the groups to which we belong and conform to the will and consensus of the majority rather than stand up against persecution and prejudice. We readily submit to the commands of individuals we perceive to have authority in the group. Whether it is our compliance to fit with what others do and say, or our obedience to follow orders, we are remarkably malleable to the pressure of the group. Our desire to be good group members seems to trump our desire to be group members who do good.

This idea is supported by two classic studies that dominate the field of compliance. The first was Stanley Milgram's obedience studies,[33] conducted at Yale in the 1960s. Here, ordinary members of the public were recruited to take part in what they thought was a study of the effects of punishment on memory. The were asked to 'teach' a student in another room to learn lists of words by punishing mistakes with increasing levels of electric shock, rising in thirty increments from an initial 15 volts to the final 450 volts. The first level was labelled 'mild' whereas the 25th level (375 volts) was labelled 'danger, severe shock'. The final two levels of 435 and 450 volts had no label other than an ominous 'XXX'. In reality, the student in the other room was a confederate of the experimenter and there were no electric shocks. The real purpose of the study was to determine how far someone would go in inflicting pain on another innocent individual when instructed to do so by an authority figure. Contrary to what the psychiatrists had predicted – they thought only about one in a hundred members of the public would obey such lethal orders – two out of three participants administered the maximum level of electric shock even though the student had been screaming and pleading to be let go. They were prepared to torture the other person to death. This is not to say most were sadists at heart; many became very distressed at the pain they were causing and yet continued to obey the orders.

The second classic study that contributed to our understanding of the way that individuals conform to group pressure is Stanford psychologist Phil Zimbardo's prison study,[34] conducted in 1971. In this mock scenario, students were recruited to take part in a two-week study of the effects of assigning the roles of prison guards and inmates in a makeshift prison built in the basement of the Stanford psychology department. The guards were told that they could not physically abuse the prisoners but they could create boredom, frustration and a sense of fear. After six days, and on the insistence of a fellow psychologist, Zimbardo abandoned the study after the guards were abusing the prisoners to such an extent that it went beyond the realms of ethical procedure. Even though they had not been given instructions to directly harm the inmates, some of the guards began to torment and torture the 'prisoners' over and beyond the original instructions. In the same way that three-year-olds were prejudiced against classmates who wore a differently coloured T-shirt, adult students took their prejudices and acted them out in violence. For Zimbardo, who interprets his study as a demonstration of the lack of personal responsibility, it was not the individuals but rather the toxic nature of the 'us'-and-'them' mentality of the situation that had created the right conditions for cruelty.

First they came . . .

When we become members of a group, we activate biases and prejudices. Even groups formed on the basis of the flip of a coin exhibit these attitudes and behaviours. We know this from the seminal work of Henri Tajfel, the former head of my department at Bristol, and subsequent studies that found the same basic automatic effects of prejudice. Before he became a psychologist, Tajfel had been a prisoner held by the Nazis during the Second World War. After the war, this experience of seeing how humans can treat and degrade their fellow man in the most appalling ways led him to spend the rest of his professional life studying the psychology of groups and how prejudice operates. Tajfel discovered that prejudice does not have to be a deep-seated, historical hatred based on politics, economics or religion. These axes to grind can aggravate any bias, but they are not an essential factor in forming prejudice. Nor does it require authority figures dictating how group members should behave. You simply have to belong to a group. Tajfel showed that simply arbitrarily assigning Bristol schoolboys into two groups by the toss of a coin produced changes in the way members of each group treated each other.[35] Even though the boys were all from the same class, those within the same group were more positive to each other but hostile to those in the other group. They went out of their way to help members of their own group, but not others.

After the war, those quick to criticize German citizens accused them of apathy because they did nothing to stop the Nazis' persecution. However, another viewpoint is one that comes from the out-group perspective. The individuals targeted were from the minorities in society so the majority did not feel threatened – it was not their problem. Initially the process was slow, during the pre-war years, so there did not appear to be a major cause for concern. Then, once the final solution was under way, people ignored what was going on.

This group mentality echoes the famous statement made after the war by the German pastor Martin Niemöller, who had spoken out about the reluctance of citizens to prevent the atrocities when he said:

> First they came for the communists,
> and I didn't speak out because I wasn't a communist.
> Then they came for the socialists,

and I didn't speak out because I wasn't a socialist.
Then they came for the trade unionists,
and I didn't speak out because I wasn't a trade unionist.
Then they came for me,
and there was no one left to speak for me.[36]

Other variations of this famous statement include the Catholics and of course the Jews, who became the focus of the 'final solution'. In addition, gypsies, homosexuals and the mentally retarded were all considered substandard humans and outcasts by the majority of German society, so it was easier to ignore their plight.

Of course, the circumstances that led up to the Holocaust are extremely complicated and there were many contributing factors. It is easy with hindsight to pass judgement on others, but the ease with which people seemed to descend into moral depravity or at least an unwillingness to help the persecuted is a testament to the power of groups. Rather than dismissing a whole nation as apathetic, anti-Semitic or even evil, it is more sensible to look for explanations that address the way people behave once they identify with a group and consider themselves to be different.

Nothing has really changed, because history repeats itself with every ethnic conflict that arises around the world. If you take our built-in tendency to be members of a tribe and the prejudices that entails, and combine it with charismatic leaders who have an agenda to coerce the group to believe that they have a legitimate grievance against an out-group, then it is easier to understand how ordinary people with no political agenda and history of racism could turn on their neighbours. Automaticity of prejudice explains how groups of otherwise peaceful citizens become violent mobs seeking out enemies of the state in hate-fuelled witch-hunts for those who have been identified as out-group members. The ease with which we take sides also explains why other countries are reluctant to get involved in these non-domestic disputes unless their interests are directly threatened. One of the most disturbing aspects of mankind is that ordinary people will turn on others who they regard as different. This is especially true when they are perceived to be in competition for resources – a bias that is exploited by political groups to stir up hatred.

These examples seem to suggest that we are all sheep, prepared to go with the crowd even when that means behaving in an immoral way. Another more plausible interpretation is that we can reinterpret our behaviour as not being bad at all, but rather for the good of the group. Even in Milgram's shocking studies, participants were more likely to comply with the instructions if they were told that it was necessary for the success of the study rather than simply told that they had no choice. Zimbardo had instructed how the guards should behave in his scientific study. It may be that these examples of extreme obedience and compliance are less about people blindly following orders but rather persuading others to believe in the importance of what they are doing. This creates a diffusion of accountability, where the individual no longer feels responsible for their actions. As British social psychologists Steve Reicher and Alex Haslam, who repeated Zimbardo's prison study in 2002, point out, 'People do great wrong, not because they are unaware of what they are doing but because they consider it to be right. This is possible because they actively identify with groups whose ideology justifies and condones the oppression and destruction of others.'[37]

Primate prejudice

It is often assumed that the prejudices that fuel group conflicts are attitudes that we have to learn. When you consider that for much of our civilization there has been constant conflict between groups of different economic, political and religious identities, then it is tempting to think that the prejudice that accompanies such conflicts must come from indoctrination. After all, national identity, political perspectives or religious beliefs are cultural inventions that we pass on to our children. And as we noted in the last chapter, we are inclined to believe what we are told. Surely we must learn to hate from others around us. However, when you look at other social animals you find evidence that prejudice is not uniquely human.

My colleague Laurie Santos at Yale wanted to know whether rhesus macaque monkeys were prejudiced.[38] Like nearly all primates, macaques live in social groups with hierarchies that are relatively stable, with dominant individuals and familial ties. The macaques Santos studies live on the beautiful Caribbean island of Cayo Santiago, a sanctuary for animals that had previously been used in US laboratories. The island is now home to around 1,000 free-ranging macaques who have formed into six distinct groups. Their social order has been well-documented, but Santos and her colleagues wanted to know if in-group members displayed evidence of prejudice against out-group members.

First they tested how individual macaques responded to static photographs of in-group and out-group members. When given a choice, macaques looked longer at the out-group individual in comparison to the in-group member. This was not because this out-group monkey was unfamiliar, because they also looked longer at a monkey who had previously spent much of its time as a member of the group before it switched allegiance to join another group. The most likely reason why they were paying extra attention to the out-group monkey was that they were being vigilant for a potential threat.

Not only do they look longer at out-group monkeys, they also associate them with unpleasant experiences. Using an ingenious technique to measure each monkey's emotional response to pictures of in- and out-group members, they found that monkeys were quicker to associate positive images of delicious fruit to pictures of in-group members and negative images of spiders with out-group members. (Like humans, monkeys hate spiders.) Not only do they aggress against out-group members, they do not like them either.

Recognizing your own group is important, but why does it feel good to belong? Humans have evolved rationality and logic to calculate the benefits of living in groups as opposed to being alone. Why do we need to feel emotions towards others as well? Feelings and emotions are two sides of the same coin. Emotions are short-lived, outward responses to an event that everyone around can read, like a sudden burst of anger or fit of laughing, but feelings are the internal lingering experiences that are not always for public consumption. We can have feelings without expressing them as emotions. They are part of our internal mental life. Without feelings, we would not be motivated to do the things we do. Feelings we get from others are some of the strongest motivations that we can have. Without feelings, there would be no point getting out of bed in the morning. Even pure logic needs feelings. When we solve a puzzle, it is not enough to know the answer – you have to feel good about it too. Why else would we bother?

It is through our social interactions that most of us find meaning in life – through the emotional experiences they generate. Pleasure, pride, excitement and love are feelings largely triggered and regulated by those around us. When we create or strive, we are not just doing it for ourselves – we seek the validation and praise of others. But others also hurt us when they cheat, lie, scold, mock, belittle or criticize. Living in groups has its ups and downs.

Social norms

Since we are social animals, it is in our collective interest not to lie, cheat or take advantage of each other in our group. This is something that good persuaders and con artists manipulate. They know that most people are kindhearted and willing to give others the benefit of the doubt when there is conflict of interest. These expectations form the basis of social norms of behaviour – what is expected by members of a group. Social norms can be so powerful that we will even apologize for something that is clearly not our fault.

Anthropologist Kate Fox deliberately bumped into commuters and jumped queues at Paddington Station in London to provoke characteristic responses that she calls the 'grammar' of social etiquette.[39] As you might have already guessed, Fox found that there is almost an automatic reflex to say 'Sorry' when we bump into strangers in the street. Failing to apologize in such a situation would be considered rude – the violation of a social norm.

We are remarkably susceptible to the power of others when it comes to conformity. A classic set of studies by American psychologist Solomon Asch in the 1950s demonstrated that individuals were also prepared to deny seeing something with their own eyes if there were enough people in the room to say otherwise.[40] He set up a situation where a real participant was part of a group with seven other confederates who were in on the true purpose of the experiment. They were told that it was a study of perception and that they had to match the length of a test line to one of three other lines. The experimenter held up a card and then went around the room, asking each person to answer aloud in turn. The real participant was among the last to be called on. The task was trivially easy. Everything was normal on the first two trials, but on the third trial, something odd happened. The confederates all began giving the same wrong answer. What did the real participant do? Results showed that three out of four of them conformed and gave the wrong answer on at least one trial.

For many decades, this research was interpreted as evidence that we comply with the group consensus. People merely said something they did not believe in order to gain social approval. It only took the presence of one other person who disagreed with the answer for the real participant to stick to their guns and give the correct answer. However, this finding has been undermined by many studies that show that even when responses are anonymous, people still go with the flow.[41]

One remarkable possibility is that people's perceptions are indeed changed by the group consensus. To get at the difference between public compliance and private acceptance, you can look at brain activation. In a recent brain-imaging study,[42] men were asked to rate photographs of 180 women for attractiveness. They were then placed in an fMRI scanner and asked to rate all the faces again, but this time they were provided with information about how each one had been rated by a group of peers. In fact, the group ratings were random. If the group said 'hot' but the participant had originally rated 'not', the participant shifted his rating higher and there was an increase in activation in two areas associated with evaluating rewards, the nucleus accumbens and the orbitofrontal cortex. Both areas light up when viewing sexually attractive faces.[43] When the group rated a face that the participant had originally thought beautiful as less attractive, there was a corresponding downward shift in his rating and brain activity.

We are so keen to fit in with the group that our behaviour can be easily manipulated. You may have noticed this with the signs and messages left for guests appearing in some of the hotels you stay at. When a Holiday Inn in Tempe, Arizona, left a variety of different message cards in their guests' bathrooms in the hopes of convincing those guests to re-use their towels rather than having them laundered every day, they discovered that the single most effective message was the one that simply read: 'Seventy-five percent of our guests use their towels more than once.'[44] This technique has recently become used to nudge people into making economic decisions that previously were imposed by the state, often raising a degree of resentment. Authorities can more easily persuade people by nudging them rather than threatening them, as a better way of influencing their behaviour.[45] When a pension fund sends out a letter saying, 'Most people are willing to invest a proportion of their earnings towards their pension . . .', the fund's managers are relying on our herd mentality to conform with the group rather than threatening us, which is less effective.

Hypocrites in the brain

How does conformity work? One answer is that when we are conforming we are avoiding the experience of discordance in our brains. It has long been known that humans need to justify their thoughts and actions; especially when they behave hypocritically. For example, if we expend a lot of effort to attain a goal to no avail, rather than accept that we have failed, we are more inclined to reframe the episode in a positive light such as 'I didn't really want that job' or 'That relationship was never going to work out'. We would rather re-evaluate the goals as negative so that we avoid discord. Aesop wrote about such 'sour grapes' in his fables as when the fox abandoned the grapes that were out of reach, dismissing them as probably inedible anyway. The reason we justify our actions is because of cognitive dissonance – the unpleasant state that arises when a person recognizes inconsistency in his or her own actions, attitudes, or beliefs. In the same way that we generally prefer truth over lies, we like to believe that we are true to ourselves.[46]

This belief means that we will frequently be disappointed in ourselves. All too often in life, we let ourselves down, which presents us with a state of dissonance – when things do not match up to our expectations. None of us is a saint – we are all flawed to a lesser or greater extent. We may cheat, lie, deceive, be economical with the truth, slack on the job, contribute less, fail to help, be hurtful, cruel or misbehave in other ways. We are often hypocritical – congratulating others through gritted teeth when we would have preferred to win the competition.

These flaws stand in direct contrast to the positive attributes we believe we possess – trustworthiness, kindness, helpfulness and generally being a good person. Very few of us are full of self-loathing or un-hypocritical. That is why there is a dissonance. Presented with the evidence of our wrongdoing, we may realize there is a contradiction. When people experience the unpleasant state of cognitive dissonance, they naturally try to alleviate it. This can be achieved by revising one's actions, attitudes or beliefs in order to restore consistency among them. So we say, 'They had it coming', 'I didn't like them in the first place anyway', or 'I always knew that they were a bad egg' – anything to reframe the situation so that whatever negative thing we have done becomes justified as a reasonable way to behave.

In one fMRI study of cognitive dissonance,[47] participants were scanned while they entertained the contradictory notion that the uncomfortable scanner environment was actually a pleasant experience. They were told that after forty-five minutes in the scanner they would be asked to rate the experience by answering questions. Half were asked to say that they actually enjoyed the experience in order to reassure a nervous participant who was waiting outside to do the study. The other half was a control group who were told that they would receive $1 each time they answered questions by saying that they enjoyed the experience. Imaging revealed that two regions were more active in the participants who had to endure the cognitive dissonance condition. These were the ACC, which detects conflicts in our thoughts and action, and the anterior insula, which registers negative emotional experiences – the same two regions that lit up during the study measuring what happens when we have to disagree with others. Not only were the ACC and the insular regions activated, but on a follow-up set of questions when there was no need to lie, the participants in the cognitive dissonance condition also rated the experience as more pleasant than the group who were paid, proving that they had indeed experienced a shift in their evaluation of the experience. In other words, they had convinced themselves that it was not such a bad experience, whereas the ones who had been paid knew they were lying for cash.

Cognitive dissonance is something that persuaders can so easily exploit. Imagine someone pushes in front of you in a queue to use a photocopier. Harvard psychologist Ellen Langer[48] found that six out of ten would not object if the person said, 'Excuse me, I only have five pages, can I use the photocopier?' Even when the apology is not intended, more than half still let the queue-jumper in front. Why is that? For one reason, most people want to avoid conflict and so do not confront the individual. They may be annoyed but not to the extent that it is worth doing something about it. Very often under these sorts of situations we will rationalize our response by reasoning that our own inconvenience is minor and thus not worth the effort. As soon as the person gives a reason such as 'Excuse me, I only have five pages, can I use the photocopier because I am in a rush?', nine out of ten do not object. By providing a reason, they have made it easier for the people waiting patiently in the queue to justify their decision to acquiesce.

We are compliant because saying 'no' is uncomfortable. Of course, there are some individuals who seem perfectly happy to barge to the front of the queue and are indifferent to others, but many of us would squirm with embarrassment. Unless, of course, we apply our own cognitive dissonance by justifying our actions, for instance by convincing ourselves that 'My needs are greater than others'. This allows us to realign our self-concept so that we do not have to entertain a contradiction that we have jumped the queue but are still really a nice person. With cognitive dissonance, we can be comfortably rude in the belief that our needs really do outweigh those of others. It is the self-deception that we discussed in the last chapter but one that applies to our whole concept of what we think we are like. Cognitive dissonance is dangerous because we can convince ourselves that we are doing the right thing even when we are not aware that we are distorting the truth. It enables us to live with our selfish behaviour and all the contradictions that entails.

Undercover racists

Most of us do not think we are hypocrites. As Aldous Huxley once wrote, 'There is probably no such thing as a conscious hypocrite.'[49] We like to think of ourselves in a positive light and very few of us would want to have all our attitudes exposed as racist, sexist or generally bigoted. And yet, despite the balanced, reasonable persona that we would like to present to the rest of the world, most of us may hold implicit ugly attitudes that are not acceptable in decent society. We know this because you can measure the level of implicit attitudes by asking participants to undertake a speeded response test where they have to match negative and positive words with different pictures.[50] It could be different races, men and women, young and old, liberals and conservatives – any of the various groups that generate stereotypes. Although most of us do not consider ourselves bigoted, the implicit attitude test reveals that we are faster to associate negative words with members of other races and positive words to members of our own group. Deep down in our unconsciousness, we have stored vast amounts of associated thoughts that reflect all the experiences and exposure to attitudes that we have encountered over our lives.

Even if we do not hold deep-seated racist attitudes, then we can still be prone to stereotypes. This has been shown in a study where white and black US adults were presented with faces of their own in-group (same race) or out-group (other race) on a computer screen.[51] When the face presented on the screen changed, they were given a painful electric shock. Eventually participants learned to associate all face changes with pain. Then the experimenters turned off the shocks to see how long it took participants to unlearn the painful association. Participants were much quicker to return to normal when the face changes were from their own race compared to faces from the other race. They took longer to become more trusting and less fearful of the other race even though they were not racist on measures taken before the test.

Does that mean that we are hard-wired to be racist irrespective of our wishes and desires? Not necessarily, because the effect was restricted to male faces and the race bias was not found in participants who had dated a member of the other race.[52] Male faces are more characteristic of threatening individuals because males are more often portrayed as aggressive. However, the racial effect can be counteracted by exposure and experience of other races. What is clear is that despite our good intentions and choices that we know we should make, biases lurk deep down in most of us that influence our decisions. These findings do not mean that we behave like this in real life, but they do reveal the problem of undercover attitudes that might surface under the right circumstances.

Judging a book by its cover

One inevitable problem of joining and identifying with groups is that we generate stereotypes that influence our judgements about and attitudes towards others. Stereotypes are assumptions that we make about all members of the same group. The problem is that stereotypes lead us to jump to conclusions that are unfair. Consider the following story about a surgeon and the unexpected shock they get one day at work:

> A father and his son were involved in a car accident in which the father was killed and the son was seriously injured. The father was pronounced dead at the scene of the accident and his body was taken to a local morgue. The son was taken by ambulance to a nearby hospital and was immediately wheeled into an emergency operating theatre. A surgeon was called. Upon arrival, and seeing the patient, the attending surgeon exclaimed, 'Oh my God, it's my son!'

How can that possibly be? If the father is dead, then how can he be the surgeon? Is there some subplot or paternity mix-up? Maybe it was the stepfather who was killed. Around half of us who read this are at a loss to explain the scenario.[53] Why are most of us so slow to realize that the surgeon is actually a woman – the boy's mother?

As Princeton's Daniel Kahneman addressed in his best-seller, *Thinking Fast, Thinking Slow*, we have two modes of thinking.[54] One is fast and automatic that occurs without intention or effort. When we make these rapid decisions about people, we quickly pigeonhole them based on the stereotypes we hold. The other type of thinking is more slow, controlled and reflective. This allows for us to consider exceptions to the rules. However, we tend to rely on the rapid process of judging people rather than defer to the more considered evaluation of others, especially when we are put on the spot. For most of us, the stereotype of a surgeon is of a white male and, having reached that decision about his identity, we find it really hard to consider that the surgeon might be female.

Rapid pigeonholing does not bode well for racial prejudice. In one speeded response task[55] adult participants earned money by 'shooting' an assailant on the screen if they were perceived to be holding a gun but punished if they were holding a camera. Of course, they made some mistakes but these were revealing. Participants were more likely to judge that a picture of a black male holding a camera showed him holding a gun instead, whereas a white male holding a gun was typically judged to be holding a camera. This was true irrespective of whether the person making the decision was white or black. Our society has become contaminated with stereotypes that we promiscuously apply out of context. This kind of stereotyped thinking is not trivial and can have fatal consequences if the one making the decision is an armed police officer.

A brain that seeks patterns in the world generates stereotypes. Our brains do this for good reasons. We build models of the world that enable us to interpret it more quickly and more efficiently. The world is also complex and confusing, so the models we build help to make sense of it. Speed, effort and efficiency mean that a stereotyping brain is going to be better adapted to deal with situations that require important decisions without the luxury of contemplative thought. Not that we have a choice. We cannot avoid building these models of the world because all experiences are filtered through the mental machinery that generates categories – summaries of our experience that chop the world up into meaningful chunks. Categorical processing is found throughout the animal kingdom, indicating that brains have evolved to seek out patterns and group them together. This happens in the brain all the way up the nervous system, from simple sensations to complex thoughts. Depending on the ecological niche a species occupies, it may only be sound and vision, but for contemplating humans it also includes judging the social groups we think others belong to and all the stereotyping that grouping entails.

Person categories refer to different classes of individuals we encounter – rich man, poor man, beggar man or thief. Each one of these categories takes many forms in terms of information, such as what they look like, how they speak, how they think and what they do. No one member is likely to tick all the boxes of the category to which they belong, but they are going to be more like each other in the same category in comparison to those from outside the category. When an individual is identified as belonging to a group, we assume they share the characteristic traits attributed to that group. This is because categories are networks of related concepts that are automatically triggered.

Another problem with pigeonholing people is that stereotypes are difficult to overcome. We accept them even when we have no evidence to either support or contradict them. We willingly accept the testimony of others because stereotypes strengthen the in-group/out-group division by attributing negative attributes to members outside our group and positive ones to our own members. We assign generalized characteristics to all members of an out-group and yet maintain that our group has much more individuality. Finally, we seek out evidence that confirms stereotypes rather than look for exceptions.[56] In a cognitive exercise known as *confirmation bias*, we select those aspects of an individual's behaviour that are consistent with our stereotype and conclude that they are typical.

Take the case of women drivers. Have you noticed how many bad women drivers there are? That, of course, is a negative stereotype that widely circulates in the West. In 2012, the mayor of Triberg, a small town in Germany, announced the opening of a new car park that had provision of a dozen 'woman only' spaces that were extra-large, well lit and near the exits.

Are women really such bad drivers? Experiments typically report superior spatial skills in males,[57] which are used to justify the claim that women are really bad at parking. However, the story is somewhat different in the real world. In the UK, the National Car Parks company conducted their own covert study[58] of 2,500 men and women using their sites and found that on average females were better at parking than males and that included the infamous reverse parking. This real-world analysis shows that women are better drivers and yet the UK Driving Standards Agency report that female drivers are more than twice as likely as males to fail their driving test on the reverse-parking manoeuvre. Are they better or not?

Females may have inferior spatial skills than males on computer lab tests, but it is probably the stereotype that women are bad at parking that is responsible for their failure on this component of the driving test. When women are reminded that males are better at maths, they perform worse in a subsequent maths test compared to women who are not primed with the stereotype.[59] The same effect was observed for African Americans who were simply reminded of their ethnicity by stating it at the beginning of an IQ test.[60] Those who wrote their race performed less well than other black students who were not reminded of the stereotype. So when it comes to parking under the scrutiny of the driving inspector, women may have a crisis of confidence and 'choke' in their performance. Simply giving women encouragement makes them more confident and improves their performance. The problem of stereotyping and why it is wrong, aside from the inequalities it creates, is that it can become a self-fulfilling prophecy.

Bad to the bone

When it comes to thinking about others, there is a real tendency to make judgements that appeal to a deeper sense of identity. As if there is something inside people that makes them who they are. This belief explains some surprising attitudes.

Would you willingly receive a heart transplant from a murderer? Under these life-or-death circumstances, I expect most people probably would, but they would be reluctant. Given a choice of organ transplantation from either a morally good person or someone who is bad, we prefer the Samaritan over the sinner.[61] It's not simply that one is evil. Rather, there is a real belief that our personality would be changed. In 1999 a British teenager had to be forcibly given a heart transplant against her will because she feared that she would be 'different' with someone else's heart.[62] She was expressing what is a common concern, namely that someone else's personality can be transferred through organ transplantation.[63] It is not uncommon for transplant patients to report psychological changes that they attribute to characteristics of the donor but there is no scientific evidence or mechanism that could explain how such a transfer could happen. There is a much more likely explanation that comes down to the way that we reason about others.

Psychological essentialism is the belief that some internal, unseen essence or force determines the common outward appearances and behaviours of category members. Even as children, we intuitively think that dogs have a 'doggie' essence, which makes them different from cats, who have a 'catty' essence. There are, of course, genetic mechanisms to explain the difference between dogs and cats, but well before mankind had made the discoveries of modern biology, people thought in terms of essences. In fact, the Greek philosopher Plato talked about the inner property that made things what they truly were. Even though individuals may not be able to say exactly what an essence is, there is a belief that there is something deep, internal and unalterable that makes an individual who they are. In this sense, it is a psychological placeholder to explain membership of one category as opposed to another.[64]

Child psychologist Susan Gelman from the University of Michigan has shown that psychological essentialism operates in young children's reasoning about many aspects of the living world.[65] By four years of age, they understand that raising a puppy in a litter of kittens will not make the puppy grow up into a cat.[66] They understand that while a stick insect may look like a stick, it is really an insect.[67] Both children and adults expect animals to maintain their identity even if external superficial features are changed. They increasingly learn to go over and beyond outward appearances when judging the true nature of things.

This explains why adults are reluctant to receive organ transplants from those that they perceive as bad. Children also develop this essentialist view. When asked about whether they would be changed by a heart transplant, six- to seven-year-olds, but not four-year-olds, thought that they would become either more or less mean and either more or less smart, depending on the psychological level of the donor.[68]

Essentialism develops well into adulthood when it comes to categorizing others into different social groups.[69] The Nazis under the guidance of Joseph Goebbels were expert at producing propaganda that demonized the persecuted as inferior, but such indoctrination was not necessary. As soon as we make a distinction between 'us' and 'them', people assume the contrasts are intrinsic, fundamental and incommensurable – they are essentially different. By adopting an essentialist perspective, we are evoking a deeper level of justification for our prejudice. We do not want to touch them. We want to keep our distance. We are making judgements about their core features because they are 'bad to the bone'. The extent to which we think of ourselves and others as possessing qualities that define who we are is a mark of our essentialist bias – a prejudice operating early in our development but one that appears to strengthen as we grow older. Psychologist Gil Diesendruck has been studying essentialist reasoning in children raised in Israel from different groups: secular Jews, Zionist Jews and Muslim Arabs. He found that by the time they are five years old, children already use category membership to make inferences about other children's personalities based on prejudice which strengthen as they grow older.[70]

Eventually, essentialism becomes enshrined in the moral codes that keep people segregated. In biological reasoning, essentialism is a useful way of categorizing the world but it is one that can be easily corrupted by those who have a prejudicial axe to grind. What is remarkable is that humans seem trip-wired to generate these distinctions and hold them without any reasonable evaluation. There is something very automatic about group membership and one of the best examples of this rapid processing is when we suddenly become aware that we have been excluded.

Social death

One day, psychologist Kip Williams from Purdue University was out walking his dog in the park when he was accidentally hit in the back with a Frisbee. He picked it up and flipped it back to the two men who had been playing, and, to his surprise, they tossed it back to Williams. Soon, he found himself enjoying an impromptu game of Frisbee with two strangers. However, this newfound friendship was short-lived. After a minute or two, the two strangers resumed passing it between themselves without any explanation or goodbyes. Williams felt hurt. He had been excluded.

What shocked Williams was his automatic reaction to this innocuous event, the pain of rejection, and how fast it kicked in. It was a humiliating experience but one that gave him a great idea. He went on to develop a computer simulation called *Cyberball*, where participants play a game in which a ball is tossed back and forth on a screen between two other playmates. Just as in his Frisbee experience, the computer includes the player for varying amounts of time and then unexpectedly excludes the player. At this point, players feel rejected. Not only that, but they feel physically hurt, which registers in the pain centres.[71] When adults played Cyberball in a brain scanner and they were excluded, their ACC, the region associated with physical pain, was activated. Their feelings were really hurt. But it also hurts to hurt others. Using the same paradigm, a recent study has shown that being forced to ostracize others is upsetting too.[72] People who were instructed to ignore others that they had just been playing with felt bad. We don't like to be made to ignore others.

Research with Cyberball reveals how easy it is to induce social pain but why should social exclusion be painful? Most pain reactions are to warn the body that damage has taken place or is about to take place. One idea is that social isolation is so damaging that we have evolved mechanisms to register when we are in danger of being ostracized.[73] This registers as pain to trigger a set of coping mechanisms to reinstate ourselves back into the social situation that threatens to expel us. As soon as it becomes clear that we are in danger of being ostracized, we activate social ingratiating strategies. We become extra helpful, going out of our way to curry favour with individuals within the group. We can become obsequious, agreeing and sucking up to others even when they are clearly in the wrong.

This is the initial response to ostracism, but if the reintegrating strategies fail, then a much more sinister, darker set of behaviours can appear. For many, the attempts to rejoin the group are replaced by aggression against the group. This aggression has been studied experimentally in a version of Milgram's shock experiment where participants believed that they were administering painful noise. Participants were asked to select an initial level of noise that ranged from 0dB to 110dB to be administered to other subjects. Prior to making their selection, they were told that increasing levels were more uncomfortable and 110dB was the maximum level. When some subjects, who in reality were experimenter confederates, rejected the real participant in a sham ostracism scenario prior to the test, he or she wreaked revenge by administering more painful sound bursts in retaliation to the others.[74] If the participant did not perceive the others as a group, they administered lower painful bursts.

Sometimes victims can be entirely innocent. In another ostracism experiment, rejected individuals spiked the food of the next participant in the study with unpleasant hot sauce even when they knew they were innocent.[75] It is the experimental equivalent of the displacement aggression when someone kicks their dog out of spite when things have gone wrong elsewhere in their lives. For many, aggression seems to be a way of getting back at an unjust world when they feel they have been injured by the thoughts and actions of others. For a few, this impulse towards revenge can be taken to the ultimate extreme.

The ultimate act of spite

> Oh the happiness I could have had mingling among you hedonists, being counted as one of you, only if you did not fuck the living shit out of me . . . Ask yourself what you did to me to have made me clean the slate.'
>
> Cho Seung-Hui's 'manifesto', describing his subsequent shooting rampage at Virginia Tech University

For many, the worst thing in the world is being rejected by others – kicked out, cut off, blackballed, sent to Coventry, unfriended. It doesn't matter how it is done. They are all ways of being ostracized. To be excluded by others is psychological death.

Exclusion is also a form of non-physical bullying, and it can sometimes have devastating consequences. In the US, the Center for Disease Control has estimated that around 4,600 children between the ages of ten and fourteen years commit suicide every year.[76] Teenage bullying is associated with depression, loneliness and suicidal thoughts.[77] Although the direct link between bullying and suicide remains to be established, contemplating killing oneself is considered a major risk factor. It is not always the physical aspects of bullying that are so detrimental, but rather the social exclusion that it usually entails. A Dutch study of nearly 4,811 schoolchildren aged between nine and thirteen years of age found that social isolation was more harmful than physical violence for both boys and girls.[78] Given the choice, teenagers would rather be hit than excluded as those who had experienced both forms of bullying reported that social aggression made them feel worse.[79] What makes these findings all the more shocking is that many teachers do not regard social exclusion as being as bad as physical bullying. In other words, not only is it difficult to monitor or police because it may go on largely unnoticed by teachers, but they can also be more tolerant of it.[80]

Rejection can also be accompanied by that other toxic thought, humiliation – ridicule and mocking by the group. No one can easily tolerate the public destruction of one's self worth, because that would make life worthless. When people feel they have been humiliated, some will wreak terrible revenge. If they do not turn the aggression in on themselves with suicide, a few will direct it back at others. They 'go postal' – a reference to the spate of US postal workers who murdered former colleagues in rampages during the 1990s.

Rampage killings are the consequences of social rejection taken to the ultimate extreme. One analysis[81] of school mass shootings such as Virginia Tech and Columbine revealed that in thirteen out of the fifteen cases, the perpetrators had been socially excluded, as illustrated so shockingly in the quote from the Virginia Tech manifesto. Others simply try to hurt society as much as possible. In the Dunblane school massacre, Thomas Hamilton targeted the most innocent victims – children – as retribution against the adults who questioned his suitability as a Scout troop leader in charge of children. In letters to the press, the BBC and even the Queen, he spelled out his resentment at his dismissal from the Scouts, which had festered for twenty-five years amidst rumours and accusations that he was a pervert, leading to his ridicule by the community. We do not yet know enough about the Sandy Hook atrocity of 2012, but gunman Adam Lanza clearly acted to inflict as much suffering as possible, and again on children. What sort of disturbed individuals could care so little about the hurt they brought to others?

One could argue that it is not that these murderers did not care about others, but rather they cared too much. They cared more about what other people thought about them than they did for the lives of their victims, their families and ultimately themselves. These atrocities were deliberate sabotage in order to be noticed. In their disturbed minds, these murderers thought they were getting even with an unjust world.

Most of us lead relatively normal lives without the extremes of ostracism and violence, but we all know what it feels like to be excluded. Even in the absence of extreme exclusion, we still lead our lives seeking the approval of others and, in doing so, maybe we all care just a little too much. Almost everything that we do is motivated by what others think, and how we are being judged.

If you ask most people about ambitions and goals, they will talk about success – something that many want but few can attain. Success is defined by what other people think. Even success in terms of material wealth and possessions has this curious aspect. We want more money to buy more of the trappings of success so that we can have status within the group. Non-material success, such as fame or infamy, is again defined by what others think. Every author writes in the hope that he or she will be read by many. Every painter wants his or her work to be appreciated. Every singer or actor wants an audience. Every politician needs support. Even the solitary rampaging gunman is motivated by what others think.

We have reached a point in our civilization where many want to be famous for the sake of just being famous, irrespective of how they go about it. There is some deep compulsion in most of us to be noticed by the group. When a small child is crying out to his parents, 'Look at me, Look at me!', they are declaring one of the fundamental needs when it comes to being human – the need for attention. That childhood urge never really goes away as we grow up into adults seeking the attention of others because they validate our existence.

The need for attention is the bittersweet twist to domestic life. Most children are raised in a nurturing environment that breeds dependency on others. Initially that dependency addresses all the physical and emotional needs that our long childhoods engender. It is a time when we learn how to become members of the groups that surround us, but when we eventually grow up to attain a level of independence, acceptance and inclusion as adults, most of us remain bound in a continual cycle of seeking approval from others. Almost everything we do is done with a view to how others will perceive us. That quest provides both the joys and the miseries of being a social animal.

EPILOGUE

A Vision of the Future?

People spend time together for a number of reasons. We may have family commitments and most of us work alongside colleagues. There are also few places on this planet where one can escape the presence of others entirely. But whether we have no choice or we actively seek out the company of others, we always prefer to be liked by the groups that we join.

Likeability depends very much on what qualities the group decides are admirable. The psychologist Richard Nisbett has argued that different cultures value different ways of behaving in groups and indeed perceive the social world differently in terms of the relatedness of individuals and groups.[1] In Eastern traditions, members of a group are interdependent and see themselves not so much as individuals but as a collective team working for the common good. This interdependence stems all the way from the family to the workplace to the whole of society. In contrast, Westerners are much more likely to consider themselves as individuals and value those who achieve success even when that comes at a cost of trampling on others on the way up to the top. Asians take great pleasure from participating in and succeeding as a group, whereas Westerners tend to take greater pride in individual achievements. Such an individualistic approach would be considerable extremely rude by most traditionalists from the East. As Nisbett points out, in Chinese there is no word for 'individualism', with the word that comes closest being one that corresponds to 'selfishness'.

Ultimately, whether collective or individualistic, validation in any culture only really exists in the minds of others. It is not enough that I believe my achievements to be a success, but rather they have to be recognized as successful by the group. This deep-seated need to be valued by the group arises because of our domesticated brains. Our success depends on acceptance from the people who inhabit our social landscape, the one that is shaped during our development. However, that landscape is now destined to change in ways that could never have been predicted.

Throughout this book, we have considered the nature of human development, both within the individual and across the evolution of our species, as a progression towards greater domestication. This I define as the related skills of coordination, cooperation and cohabitation with others, based on what is considered acceptable behaviour. Other animals have some of these attributes of coexistence as well, but no other animal has taken domestication to the extent that we have. Rudiments of coordination, cooperation and cohabitation must have existed in our species from the very beginning, when hominids became socially dependent on one another hundreds of thousands of years ago. Each of these social skills required a brain capable of perceiving others in terms of who they are, what they want, what they are thinking and, in particular, what are they thinking about me? Coordination enabled many to work together for more than could be achieved by the individual working alone. Cooperation was a spur to help each other on the understanding that there would be reciprocal benefits down the road. Cohabitation provided safety and security in numbers as well as the change in our species from a nomadic to a sedentary domesticated life.

So what does the future hold for this domesticated life? We are currently living through one of the most transformative periods in human history. Every so often a new technology comes along that changes the way we behave. The raft was a significant invention as it enabled groups of early humans to migrate across oceans to new territories. The plough played a critical role in the birth of agriculture that spurred the transition from a nomadic lifestyle to a sedentary existence that is the basis for modern domestic living. Gunpowder and steel changed the way certain groups conquered and subjugated other peoples.[2] Printing enabled humans to spread knowledge and create our education systems.

The invention of the Internet will go down as another significant milestone in the evolution of human civilization. It is an unprecedented system for exchanging information and conducting business, but it is the social revolution the Internet has created that is probably the most unforeseen consequence of this technology. Not so long ago, we may have spent most of our time actually in the presence of others, but that was before the Internet infiltrated almost every household in the West. With around 1.73 billion subscribers, nearly one in four people on the planet currently uses social networking sites (SNS), with a forecast of 2.55 billion users by 2017.[3] It seems inevitable that eventually the majority of the human race will be online, engaging in social exchanges. For the first time in the history of our species, each one of us has the potential capability to interact with anyone else on the planet in real time but within a virtual environment. We have come a long way from the small band of early hominids living in small numbers on the African savannah, exchanging gossip within our group. The social skills that we evolved for interacting with each other are now brought to bear in situations where we communicate with not just a handful but hundreds and sometimes thousands of others over vast distances in any time zone from the comfort of our homes.

Still, for many people, the Internet is something to be feared. As with any new technologies, from the printing press to the radio, there is always anxiety that change is not good because outcomes are unpredictable. *Technopanic* is a term that captures the fears about the way the Internet is changing the way humans behave.[4] The British neuroscientist Susan Greenfield warns that the Internet is wreaking irrevocable damage on our children's developing brains because they are not using the communication skills that were honed over evolution.[5] Philip Zimbardo, the psychologist famous for his Stanford Prison study, tells us that the widespread availability of online pornography is leading to the 'demise of guys', who are now unable to inhibit their sexual compulsions and failing to learn how to interact appropriately with women.[6] In the UK in 2013, the Coalition government looked into regulating Internet searches for sexual content despite the lack of any clear evidence that it is a problem.[7] We read of extreme cases of online addiction to virtual communities or gaming where individuals play for days on end, sometimes leading to the deaths of themselves and even their own children that they have neglected.[8]

All of these sensationalist headlines seem to be hysterical technopanic based on scant evidence or simply anecdotal reports. There has not been enough time to conduct the proper analysis to fully test the claims in this fast-changing world of information technology. However, one only has to consider world poverty or climate change to realize that Internet addiction is one of the least of our worries. But all of us, especially those who remember the pre-Internet days, cannot fail to be astounded by the blistering pace of change and the uncertain future it will create. It is easy to appreciate why those fearful of change consider the Internet as a force for evil.

As a parent of two teenage girls, I am less concerned by the threats that the Internet seems to pose for the future of our children. I do not believe that the Internet will doom them to compassionless relationships. Rather, as I watch them use the Internet for social networking, it is clear that they are enjoying much greater freedom and an exposure to a greater diversity of ideas than was previously possible. No wonder that oppressive regimes try to suppress and control the Internet to prevent their own citizens getting the 'wrong' ideas.

For all its benefits, however, it would be foolhardy not to consider how the Internet will change the way we interact and the potential problems this may entail. Humans bring a legacy of our evolutionary past to this brave new world where social interactions in the future are likely to be very different. Our species was not adapted to this digital environment, and how we behave will probably change as a result of this complex interaction between our biology, psychology and technology in ways that we are still trying to unravel.

To begin, rather than seeking approval from a few select friends in person, it is clear that we will increasingly be influenced by the group. For example, SNS can generate appreciation and validation from large numbers on the Internet. This is especially true of Twitter, which is effectively open texting to the world. Twitter provides the opportunity to monitor, and be monitored by, anyone almost completely anonymously. Even though these interactions are virtual, studies show that acceptance and rejection can be just as emotionally charged on the Internet as such encounters in real life.[9]

So what are we doing on these SNS? The short answer is talking about ourselves. During normal conversation, we spend about 30–40 per cent of the time talking about ourselves, which according to brain-imaging studies makes us feel good.[10] The brain regions associated with rewards and pleasures are activated when describing our experiences. On the Internet we take this self-obsession to the extreme. Over 80 per cent of the posts on SNS are about the poster. Already, we seem to be hooked. A study of over 1,000 Swedish Facebook subscribers found that the average user logs on to the site six times per day, spending an average seventy-five minutes – women more than men.[11] One in four report that they feel uneasy when they cannot access SNS. We love to talk about ourselves, which is why SNS are such an enticing opportunity. Here we seem oblivious to the social barriers or restraints about how much you can go on and on about yourself.

When SNS first appeared they offered the opportunity to enable people to connect and keep in touch – something that was increasingly compromised in a society where people were living busy lives that often forced them to move on to new pastures. SNS offer a chance for those who are isolated to make new friends or keep in contact with others that have moved on. However, true friends are few and making acquaintances on SNS hardly offers the same experience. Moreover, there are drawbacks to exposing yourself to a vast audience with whom there is no direct face-to-face interaction and the strength of friendship is weak.

Ironically, one major danger of too many friends may be damage to self-esteem. Contrary to expectations, SNS do not help those with low self-esteem by giving them a platform to express themselves without the pressure of social anxiety that real encounters can generate. Rather, they amplify their problems. The trouble with individuals with low self-esteem is that they talk more openly about the negative aspects of their lives and personality, which are not appealing topics of conversation for those on the Internet. The irony is that they may feel more secure in revealing things about themselves on SNS, but the rest of us do not want to hear how bad their lives are, which leads us to push them away.[12]

We are so self-obsessed that we tend to only pay attention to the information that relates to us. When you accumulate large numbers of friends on SNS, this is a tangible measure of popularity. When someone of high status, such as a celebrity, follows you on Twitter, then you can bask in their reflected glory as someone worthy of their attention.[13] The whole SNS phenomenon may have originally been intended to share experiences and opinions, but has become a mechanism for narcissism.

'Selfies' are the latest craze – posting pictures of ourselves so that others can look at us. Even at the memorial ceremony for Nelson Mandela, heads of state were taking selfies. A 2013 poll by Samsung, one of the manufacturers of the ubiquitous camera phones, revealed that selfies accounted for 30 per cent of pictures taken by eighteen- to twenty-four-year-olds.[14] On Facebook, the largest of the SNS, its users click 'Like' 2.7 billion times and share 300 million photographs per day.[15] This can lead to an inflated sense of self-esteem, boosted by all the 'Like's, positive comments or recommendations that others bestow upon us. This concern for what others think can also lead to extremism caused by polarization. If we only listen to those who agree with us, then the tendency will be to become more certain of our opinions, intolerant to criticism or, worse still, to become more radical in order to be seen to be more forthright.[16]

Some individuals use SNS to bully and harass. Already there has been a spate of teenage suicides attributed to cyberbullying, though it is not clear whether this reflects a significant increase in this troubled age group.[17] We can also become indignant and intolerant of others more easily on the Internet than in a real-life encounter. Somehow, like the road rage that we have all witnessed or experienced when drivers are isolated in their cars, people behave differently when they are not in a face-to-face situation. The Internet is a place to vent anger or take revenge on others from the comfort of our own home. Nobody likes to be criticized, but criticism can be particularly painful on the Internet because it is such a public arena. What were once local and personal grievances that could be settled by a measured response or gesture can escalate into dramas broadcast to the world to reveal the sense of injustice the injured party feels.

For many, it seems that we have gone too far with the Internet, but the revolution in social behaviour has only really just begun. Increasingly, everything we do and everywhere we go is being shared with others. Every piece of software, every purchase, every choice we make is no longer an individual secret but a valuable piece of data worth sharing. Vint Cerf, one of the fathers of the Internet and a digital visionary, predicts that soon our clothing will be capable of relaying information up to the net.[18] Already most of us possess smartphones that keep wanting to share information about where we are, what we are doing and what we like. We no longer have to pay for many services and applications; we simply have to let others know that we are using them across the SNS. This is because businesses know that social information is the key to success. The individual choices that we think are ours are being used to inform the group, which in turn is being used to influence our choices in a vast online social experiment of conformation bias.

We really do not have a choice. It is becoming impossible to be anonymous. Just about all of us in the West are dependent on the goods and services of others that we have to purchase. In the past this could be done in anonymous transactions, but eventually hard cash will disappear and so will our capacity to remain elusive. Transactions will all become digital and, with that, your identity will be used to catalogue your activity.

As we increasingly go online, algorithms that keep track of our search activity will seek to anticipate what we want and will tailor the choices made available to us by only sending us the information that best fits our search requirements. Marketing companies want to personalize their offerings to each of us. The problem is that this tailoring creates 'filter bubbles', where information deemed less relevant is shielded from us.[19] In the drive to produce increasingly personalized functionality on websites by monitoring our activity, there is currently a big data gold rush where companies such as Google and Facebook are collecting personal information that can be sold on to marketing companies. With these vast databases, the collective opinion of the groups to which we belong will not only begin to shape the decisions we make but limit the options we are offered in an attempt to optimize the choices we have to make.

It is all meant to make life more convenient but it will also make it more conventional. Where once Western independence and Eastern interdependence existed as geographically separated cultural social norms, the global reach of the Internet and the way our choices are being shaped and curtailed by the behaviour of the group threatens our capacity to maintain unique identity and privacy.

It is happening right now. Soon we may have no choice at all. The virtual world is spilling over into the real world.

The technology already exists with Google glasses, where the wearer can upload live sights and sounds to the Internet for others to watch (though Google claims this will be regulated). Soon the technology will be so small as to be invisible, so that you will not be aware that you are being watched. People will never be sure that they are truly alone or having a private conversation. This was the prediction in George Orwell's *1984* and became the inspiration for the hit reality TV show *Big Brother* in the last decade. Contestants willingly participated in *Big Brother* for the prospect of fame and celebrity status, but were really selected by producers because they were the most colourful and often dysfunctional individuals – a modern version of the Victorian freak shows described in Chapter 3. Still, applicants to those shows made the choice to be watched, and we as the audience made the choice to watch. Today, the Internet threatens to put all of us under surveillance whether we like it or not.

When modern humans first left Africa some 60–70,000 years ago, they had the necessary social expertise to live together, and to seek out new territories as the vast Northern ice sheets began to recede. They communicated and co-operated with brains that were able to pass knowledge on to each new generation. They developed feelings, behaviours and thoughts that were evolved to keep them connected. At the end of the last Ice Age 20,000 years ago, humans began to settle down and change from a nomadic lifestyle to one where they learned to cultivate crops and rear animals.

Throughout our evolution, domestication has provided strength in numbers for the individual, but that same domestication that enabled us to live so well together now threatens to eradicate the individual. We have become so dependent on others that few of us could be self-sufficient and there is little sign that this co-dependency has reached its peak. Co-dependency provides an easier life and this increasingly relies on information technology. However, we seem to be largely unaware that this innovation is being used to both monitor and shape the way we live.

Our domesticated brain allowed us to become animals that thrive by living together in groups, but with our technological advances, the size of that group has almost become unlimited by geography or time zones. One wonders whether an ever-expanding group will eventually subsume us. Maybe there will always be tensions to resist the pull of the crowd. One can imagine a future of continual cultural conflict as we try to maintain our group identities in the face of increasing integration. That said, losing our group identities and the prejudices that separate us may be the necessary solution that enables humans to coordinate, cooperate and cohabit on a planet of limited resources. When we start thinking and acting as a group on a global scale, we will be better suited to cope with many of the problems our species faces – population growth, food shortages, deforestation, pandemics and maybe even climate change.

ACKNOWLEDGEMENTS

This is my third popular-science book and it does not get any easier. I am lucky to have the best agent in Robert Kirby. He really is the best, as he won a prize to say so. Robert's unfailing support and enthusiasm make me want to call him up all the time just to hear his reassuring voice of wisdom. Thank you also to my editor, Laura Stickney, who gave me the freedom to write the book I needed to. I hope it helps to put Pelican back where it belongs in the hearts and minds of the general public.

The ideas in this book are very much influenced by many other scholars from whom I have borrowed liberally. However, special mention should be made of a number of them, including Paul Bloom, Brian Hare, Philippe Rochat and especially Michael Tomasello, whose work on cooperation can be found throughout the text. I am particularly indebted to Cristine Legare, who not only read the whole book but provided invaluable guidance and recommendations. I also need to thank my students and colleagues who have read sections or given me food for thought. These include Sara Baker, Shiri Einav, Iain Gilchrist, Nathalia Gjersoe, Kiley Hamlin, Pat Kanngieser, Kate Longstaffe, Marcus Munafo, Laurie Santos and Sandra Weltzien. Of course, my writing would not be possible without the support of my long-suffering family, who have had to put up with my unreasonable behaviour over the years.

This book is dedicated to my mother, Loyale Hood, who domesticated me, or at least made a good attempt at it. Thank you, mum.

Notes

PREFACE

1. Kathleen McAuliffe (2010), 'If Modern Humans Are So Smart, Why Are Our Brains Shrinking?', *Discover* magazine, September 2010. http://discovermagazine.com/2010/sep/25-modern-humans-smart-why-brain-shrinking#.UdGTQxxYdNo
2. D. H. Bailey and D. C. Geary (2009), 'Hominid brain evolution: Testing climactic, ecological, and social competition models', *Human Nature*, 20, 67–79.
3. D. C. T. Kruska (2005), 'On the evolutionary significance of encephalization in some eutherian mammals: effects of adaptive radiation, domestication, and feralization', *Brain, Behavior and Evolution*, 65, 73–108.
4. Claudio J. Bidau (2009), 'Domestication through the centuries: Darwin's ideas and Dmitry Belyaev's long-term experiment in silver foxes', *Gayana* 73 (Suplemento), 55–72.
5. Hilleke Pol et al. (2006), 'Changing your sex changes your brain: influences of testosterone and estrogen on adult human brain structure', *European Journal of Endocrinology*, 155, S107–S114.
6. K. Soproni, Á. Miklósi, J. Topál and V. Csányi (2001), 'Comprehension of human communicative signs in pet dogs (*Canis familiaris*)', *Journal of Comparative Psychology*, 115, 122–6.
7. L. N. Trut (1999), 'Early canid domestication: the farm fox experiment', *American Scientist* 87 (March–April), 160–9.
8. Brian Hare, Victoria Wobber and Richard Wrangham (2012), 'The self-domestication hypothesis: evolution of bonobo psychology is due to selection against aggression', *Animal Behavior*, 83, 573–85.
9. B. Hare (2007), 'From Nonhuman to Human Mind. What Changed and Why', *Current Directions in Psychological Science*, 16, 60–4.
10. Adam Brumm, Gitte M. Jensen, Gert D. van den Bergh, Michael J. Morwood, Iwan Kurniawan, Fachroel Aziz and Michael Storey (2010), 'Hominins on Flores, Indonesia, by one million years ago', *Nature*, 464, 748–52.

11. Alan Simmons (2012), 'Mediterranean island voyages', *Science*, 338, 895–7.
12. Adam Powell, Stephen Shennan and Mark G. Thomas (2009), 'Late Pleistocene demography and the appearance of modern human behavior', *Science*, 324, 1298–1301.
13. H. Zheng, S. Yan, Z. Qin and L. Jin (2012), 'MtDNA analysis of global populations support that major population expansions began before Neolithic Time', *Scientific Reports*, 745; DOI:10.1038/srep00745.

CHAPTER 1

1. Dan Wolpert opens with this question in his TED talk: http://www.ted.com/talks/daniel_wolpert_the_real_reason_for_brains.html
2. The example of the sea squirt is given by a variety of authors but most notably Rodolfo R. Llinás (2001), *I of the Vortex: From Neurons to Self*, MIT Press.
3. F. de Waal (2013), *The Bonobo and the Atheist: In Search of Humanism Among the Primates*, W. W. Norton & Company.
4. Jane Goodall (1986), *The Chimpanzees of the Gombe: Patterns of Behavior*, Cambridge: The Belknap Press of Harvard University Press.
5. M. Nakamichi, E. Kato, Y. Kojima and N. Itoigawa (1998), 'Carrying and washing of grass roots by free-ranging Japanese macaques at Katsuyama', Folia Primatologica *International Journal of Primatology*, 69, 35–40.
6. Lydia V. Luncz, Roger Mundry and Christophe Boesch (2012), 'Evidence for cultural differences between neighboring chimpanzee communities', *Current Biology*, 22, 922–6.
7. Richard Dawkins (1976), *The Selfish Gene*, Oxford University Press.
8. Richard Dawkins (1996), *The Blind Watchmaker: Why the Evidence of Evolution Reveals a Universe without Design*, New York: Norton & Company.
9. M. E. J. Newman and R. G. Palmer (1999), 'Models of Extinction: A Review', Santa Fe Institute working paper, http://www.santafe.edu/media/workingpapers/99-08-061.pdf.
10. The 'environmental complexity hypothesis' argues that one of the driving forces for developing intelligence supported by larger brains was the need to adapt to variable environments.

    Grove, M. (2011), 'Change and variability in Plio-Pleistocene climates: Modelling the hominin response', *Journal of Archaeological Science*, 38, 3038–47.

11. X. H. Zhu, H. Qiao, F. Du, Q. Xiong, X. Liu, X. Zhang, K. Ugurbil and W. Chen (2012), 'Quantitative imaging of energy expenditure in human brain', *Neuroimage*, 60, 2107–17.
12. D. Attwell and S. B. Laughlin (2001), 'An energy budget for signaling in the grey matter of the brain', *Journal of Cerebral Blood Flow and Metabolism*, 21, 1133–45.
13. L. Marino (1998), 'A comparison of encephalization between odontocete cetaceans and anthropoid primates', *Brain, Behavior, and Evolution*, 51, 230–8.
14. I. Loudon (1986), Deaths in childbed from the eighteenth century to 1935', *Medical History*, 30, 1–41.
15. J. DeSilva and J. Lesnik (2006), 'Chimpanzee neonatal brain size: Implications for brain growth in Homo erectus', *Journal of Human Evolution*, 51, 207–12.
16. A. Portmann (1969), 'Biologische Fragmente zu einer Lehre vom Menschen [A Zoologist Looks at Humankind] (Schwabe, Basel, Germany); trans. J. Schaefer (1990), New York: Columbia University Press.
17. C. D. Bluestone (2005), 'Humans are born too soon: impact on pediatric otolaryngology', *International Journal of Pediatric Otorhinolaryngology*, 69, 1–8.
18. J. H. Kaas (2005), 'From mice to men: the evolution of the large, complex human brain', *Journal of Bioscience*, 30, 155–65.
19. Holly M. Dunsworth, Anna G. Warrener, Terrence Deacon, Peter T. Ellison and Herman Pontzer (2012), 'Metabolic hypothesis for human altriciality', *Proceedings of the National Academy of Sciences USA*, 109, 15212–16.
20. R. D. Martin (1996), 'Scaling of the mammalian brain: The maternal energy hypothesis', *News in Physiological Science*, 11, 149–56.
21. K. R. Rosenberg and W. R. Trevathan (2003), 'The Evolution of Human Birth', *Scientific American,* May: 80–85.
22. Milton, Katharine (2000), 'Diet and Primate Evolution' in Alan Goodman, Darna Dufour and Gretel Pelto (eds), *Nutritional Anthropology: Biocultural Perspectives on Food and Nutrition*, Mountain View, CA: Mayfield Publishing Company, 46–54.
23. G. T. Frost (1980), 'Tool behavior and the origins of laterality', *Journal of Human Evolution*, 9, 447–59.
24. John Allen (2009), *The Lives of the Brain: Human Evolution and the Organ of Mind,* Belknap Harvard
25. The Leakeys reported the discovery of fossils at Koobi Fora, near Lake Turkana in Kenya, that indicate that three separate hominid species coexisted as early as 2 million years ago. M.G. Leakey, F. Spoor, M.C. Dean, C.S. Feibel, S.C. Antón, C. Kiarie and L.N. Leakey (2012), 'New fossils from Koobi

Fora in northern Kenya confirm taxonomic diversity in early *Homo*', *Nature*, 488, 201–204.

However, recent discoveries in the Georgian village of Dmanisi of a variety of skull shapes from *Homo erectus* dating from 2 million years ago suggest much less evidence for distinct species of hominids evolving in Africa based on different skull shapes.

David Lordkipanidze, Marcia S. Ponce de León, Ann Margvelashvili, Yoel Rak, G. Philip Rightmire, Abesalom Vekua and Christoph P. E. Zollikofer (2013), 'A Complete Skull from Dmanisi, Georgia, and the Evolutionary Biology of Early Homo', *Science*, Vol. 342 no. 6156, 326–31.

26. I. McDougall, F. H. Brown and J. G. Fleagle (2005), 'Stratigraphic placement and age of modern humans from Kibish, Ethiopia', *Nature*, 433, 733–6.
27. R. L. Cann, M. Stoneking and A. C. Wilson (1987), 'Mitochondrial DNA and human evolution', *Nature*, 325, 31–6.
28. Annalee Newitz (2013), A long anthropological debate may be on the cusp of resolution, http://io9.com/a-long-anthropological-debate-may-be-on-the-cusp-of-res-512864731 (Interview with Ian Tattersall)
29. University of Montreal (2011, July 18), 'Non-Africans are part Neanderthal, genetic research shows', *Science Daily*. Retrieved July 4, 2013, from http://www.sciencedaily.com/releases/2011/07/110718085329.htm
30. R. I. M. Dunbar and S. Shultz (2007), 'Evolution in the social brain', *Science*, 317, 1344–7.
31. J. B. Silk (2007), 'Social components of fitness in primate groups', *Science*, 317, 1347–51.
32. M. Gutison et al. (2012), 'Derived vocalizations of geladas (Theropithecus gelada) and the evolution of vocal complexity in primates', *Philosophical Transactions of the Royal Society of Biological Sciences*, 367, 1847–59.
33. Nicola Clayton (2012), 'Corvid cognition: Feathered apes', *Nature*, 484, 453–4.
34. Chris Stringer (2011), *The Origin of our Species*, London: Allen Lane.
35. H. Zheng, S. Yan, Z. Qin and L. Jin (2012), 'MtDNA analysis of global populations support that major population expansions began before Neolithic Time', Sci. Rep. 2, 745; DOI:10.1038/srep00745.
36. S. Mithen, (1996), *The Prehistory of the Mind: A Search for the Origins of Art, Religion and Science*, London: Thames and Hudson.
37. Nicholas Humphrey (1983), *Consciousness Regained*, Oxford University Press.
38. Frans de Waal (2007), *Chimpanzee Politics: Power and Sex Among Apes* (25th Anniversary ed.), Baltimore, MD: JHU Press.
39. D. G. Premack and G. Woodruff (1978), 'Does the chimpanzee have a theory of mind?', *Behavioral and Brain Sciences*, 1, 515–26.

40. Steven Pinker (1994), *The Language Instinct*, New York: Morrow.
41. R. I. M. Dunbar (1996), *Grooming, Gossip and the Evolution of Language*, London: Faber & Faber.
42. S. Pinker and P. Bloom (1990), 'Natural language and natural selection', *Behavioral and Brain Sciences*, 13, 707–84.
43. A. A. Ghazanfar and D. Rendall (2008), 'Evolution of human vocal production', *Current Biology*, 18, 457–60.
44. K. S. Lashley (1951), 'The problem of serial order in behavior' in L. A. Jefress (ed.), *Cerebral mechanisms in behavior*, New York: Wiley, 112–46.
45. Noam Chomsky (1986), *Knowledge of Language: Its Nature, Origin and Use*, New York: Praeger.
46. Steve Pinker (1994), *The Language Instinct: How the Mind Creates Language*, New York: William Morrow & Co.
47. Leda Cosmides and John Tooby (1994), 'Origins of domain specificity: The evolution of functional organization' in L. Hirshfeld and S. A. Gelman (eds), *Mapping the Mind: Domain specificity in cognition and culture*, New York: Cambridge University Press.
48. Maciej Chudek, Patricia Brosseau-Laird, Susan Birch and Joseph Henrich (2013), 'Culture-Gene Coevolutionary Theory and Children's Selective Social Learning' in M. R. Banaji and S. A. Gelman (eds), *Navigating the Social World. What Infants, Children, and Other Species Can Teach Us*, New York: Oxford University Press.
49. Mike Tomasello (2009), *Why We Cooperate*, Boston: Boston Review.
50. Felix Warneken, Brian Hare, Alicia P. Melis, Daniel Hanus and Michael Tomasello (2007), 'Spontaneous altruism by chimpanzees and young children', PLoS Biol 5(7): e184. doi:10.1371/journal.pbio.0050184.
51. Judith M. Burkart, Ernst Fehr, Charles Efferson and Carel P. van Schaik (2007), 'Other-regarding preferences in a non-human primate: Common marmosets provision food altruistically', *Proceedings of the National Association of Sciences*, 104, 19762–6.

CHAPTER 2

1. John Locke, (1690), *An Essay Concerning Human Understanding*, New York: E. P. Dutton, 1947.
2. William James (1890), *Principles of Psychology*, New York: Henry Holt.

3. Immanual Kant (1781), *Critique of Pure Reason*, trans. J. M. D. Meiklejohn, The Electronic Classics Series, ed. Jim Manis, PSU-Hazleton, Hazleton, PA.
4. J. M. Fuster (2003), *Cortex and Mind*, New York: Oxford University Press
5. F. A. C. Azevedo et al. (2009), 'Equal numbers of neuronal and nonneuronal cells make the human brain an isometrically scaled-up primate brain', *Journal of Comparative Neurology*, 513, 532–541.

   This is the most recent analysis of the human neural architecture. They estimated that there were 85 billion non-neuronal cells and 86 billion neuronal cells.
6. R. C. Knickmeyer, S. Gouttard, C. Kang, D. Evans, K. Wilber, J. K. Smith et al. (2008), 'A structural MRI study of human brain development from birth to 2 years', *Journal of Neuroscience*, 28, 12176–82.
7. Gregory Z. Tau and Bradley S. Peterson (2010), 'Normal development of brain circuits', *Neuropsychopharmacology Reviews*, 35, 147–68.
8. David A. Drachman (2005), 'Do we have brain to spare?', *Neurology*, 64, 2004–5.
9. This is a paraphrase of Donald Hebb's rules of neuronal learning and synaptic plasticity.

   D. O. Hebb (1949), *The Organization of Behavior*, New York: Wiley & Sons.
10. Elizabeth S. Spelke, (2000), 'Core knowledge', *American Psychologist*, 55, 1233–43.
11. Valerie A. Kuhlmeier, Paul Bloom and Karen Wynn (2004), 'Do 5-month-old infants see humans as material objects?', *Cognition*, 94, 95–103.
12. Aina Puce and David Perrett (2003), 'Electrophysiology and brain imaging of biological motion', *Phil. Trans. R. Soc. Lond. B* (2003) 358, 435–45.
13. Virginia Slaughter, Michelle Heron-Delaney and Tamara Christie (2012), 'Developing expertise in human body perception' in V. Slaughter and C.A Brownell (eds), *Early Development of Body Representations. Cambridge Studies in Cognitive and Perceptual Development* 13, Cambridge, UK: Cambridge University Press, 207–26.
14. F. Simion, L. Regolin and H. Bulf (2008), 'A predisposition for biological motion in the newborn baby', *Proceedings of the National Academy of Sciences* (USA), 105, 809–13.
15. A. J. DeCasper and M. J. Spence (1986), 'Prenatal maternal speech influences newborns' perception of speech sounds', *Infant Behavior & Development*, 9, 133–50.
16. Aidan Macfarlane (1975), 'Olfaction in the development of social preferences in the human neonate' in A. Macfarlane (ed.), *Parent-Infant Interactions*, Amsterdam: Elsevier, pp. 103–17.

17. Dare A. Baldwin (2013), 'Redescribing Action' in M. R. Banaji and S. A. Gelman (eds), *Navigating the Social World. What Infants, Children, and Other Species Can Teach Us*, New York: Oxford University Press.
18. D. A. Baldwin, J. A. Baird, M. Saylor and M. A. Clark (2001), 'Infants parse dynamic human action', *Child Development*, 72, 708–17.
19. Margaret Legerstee (1992), 'A review of the animate-inanimate distinction in infancy. Implications for models of social and cognitive knowing', *Early Development and Parenting*, 1, 59–67.
20. F. Heider and M. Simmel (1944), 'An experimental study of apparent behavior', *American Journal of Psychology*, 57, 243–59.
21. Dan C. Dennett (1971), 'Intentional systems', *Journal of Philosophy*, 68, 87–106.
22. Val Kuhlmeier, Karen Wynn and Paul Bloom (2003), 'Attribution of dispositional states by 12-month-olds', *Psychological Science*, 14, 402–8.
23. J. Kiley Hamlin, Karen Wynn and Paul Bloom (2010), 'Three-month-olds show a negativity bias in their social evaluations', *Developmental Science*, 13, 923–9.
24. A. L. Yarbus (1967), *Eye Movements and Vision* (trans. B. Haigh), New York: Plenum Press.
25. Stewart Guthrie (1993), *Faces in the Clouds: A New Theory of Religion*, Oxford University Press.
26. Nancy Kanwisher, Josh McDermott and Marvin M. Chun (1997), 'The Fusiform Face Area: A module in human extrastriate cortex specialized for face perception', *Journal of Neuroscience*, 17, 4302–11.
27. Michael Argyle and Janet Dean (1965), 'Eye-contact, distance and affiliation', *Sociometry*, 28, 289–304.
28. A. Frischen, A. P. Bayliss, S. P. Tipper (2007), 'Gaze cueing of attention: visual attention, social cognition, and individual differences', *Psychological Bulletin*, 133, 694–724.
29. Bruce Hood, Doug Willen and Jon Driver (1998), 'An eye direction detector triggers shifts of visual attention in human infants', *Psychological Science*, 9, 53–6.
30. M. Von Grünau and C. Anston (1995), 'The detection of gaze direction: a stare-in-the-crowd effect', *Perception*, 24, 1297–1313.
31. Reginald B. J. Adams, Heather L. Gordon, Abigail A. Baird, Nalini Ambady and Robert E. Kleck (2003), 'Effects of gaze on amygdala sensitivity to anger and fear faces', *Science*, 300, 1536.
32. Teresa Farroni, Gergely Csibra, Francesca Simion and Mark H. Johnson (2002), 'Eye contact detection in humans from birth', *Proceedings of the National Academy of Sciences USA*, 99, 9602–5.

33. S. M. J. Hains and D. W. Muir (1996), 'Effects of stimulus contingency in infant–adult interactions', *Infant Behavior & Development*, 19, 49–61.
34. M. Argyle and M. Cook, *Gaze and Mutual Gaze* (1976), Cambridge University Press.
35. H. Akechi, A. Senju, H. Uibo, Y. Kikuchi, T. Hasegawa et al. (2013), 'Attention to Eye Contact in the West and East: Autonomic Responses and Evaluative Ratings', PLoS ONE 8(3): e59312. doi:10.1371/journal.pone.0059312.
36. J. Kellerman, J. Lewis and J. D. Laird (1989), 'Looking and loving: The effects of mutual gaze on feelings of romantic love', *Journal of Research in Personality*, 23, 145–61.
37. E. Nurmsoo, S. Einav and B. M. Hood (2012), 'Best friends: children use mutual gaze to identify friendships in others', *Developmental Science*, 15, 417–25.
38. M. Bateson, D. Nettle and G. Roberts (2006), 'Cues of being watched enhance cooperation in a real-world setting', *Biology Letters*, 2, 412–14.

    D. Francey and R. Bergmüller (2012), Images of eyes enhance investments in a real-life public good. PLoS ONE 7, e37397.

    Kate L. Powell, Gilbert Roberts and Daniel Nettle (2012), 'Eye images increase charitable donations: Evidence from an opportunistic field experiment in a supermarket', *Ethology*, 118, 1–6.

    M. Ernest-Jones, D. Nettle and M. Bateson (2011), 'Effects of eye images on everyday cooperative behavior: a field experiment', *Evol. Hum. Behav.* 32, 172–8.
39. Mike Tomasello (2009), 'Why We Cooperate', *Boston Review*.
40. M. Tomasello and M. J. Farrar (1986), 'Joint attention and early language', *Child Development*, 57, 1454–63.
41. G. Butterworth (2003), 'Pointing is the royal road to language for babies' in S. Kita (ed.), *Pointing: Where language, culture, and cognition meet*, Mahwah, NJ: Erlbaum, pp. 9–33.
42. Others believe that apes share all the same communicative gestures and joint attention as humans. David A. Leavens (2012), 'Joint attention: twelve myths' in *Joint attention: New developments in Psychology, Philosophy of Mind, and Social Neuroscience*, Cambridge, Mass.: MIT Press, pp. 43–72.
43. Anne Fernald and T. Simon (1984), 'Expanded intonation contours in mothers' speech to newborns', *Developmental Psychology*, 20, 104–13.
44. Andrew N. Meltzoff and Rechele Brooks (2001), '"Like me" as a building block for understanding other minds: bodily acts, attention, and intention' in Betram F. Malle and Dare Baldwin (eds), *Intentions and Intentionality: Foundations of Social Cognition*, Cambridge, Mass.: MIT Press, 171–91.

45. Rod Parker-Rees (2007), 'Liking to be liked: imitation, familiarity and pedagogy in the first years of life', *Early Years*, 27, 3–17.
46. Andrew N. Meltzoff (1995), 'Apprehending the intentions of others. Re-enactment of intended acts by 18-month-old children', *Developmental Psychology*, 31, 838–50.
47. G. Gergely, H. Bekkering and I. Kiraly (2002), 'Rational imitation in preverbal infants', *Nature*, 415, 755.
48. V. Horner and A. Whiten (2005), 'Causal knowledge and imitation/emulation switching in chimpanzees (Pan troglodytes) and children (Homo sapiens)', *Animal Cognition*, 8, 164–81.
49. Derek E. Lyons, Andrew G. Young, and Frank C. Keil (2007), 'The hidden structure of over imitation', *Proceedings of the National Academy*, 104, 19751–6.
50. P. A. Herrmann, C. H. Legare, P. L. Harris and H. Whitehouse (2013), 'Stick to the script: The effect of witnessing multiple actors on children's imitation', *Cognition*, 129, 536–43.
51. C. H. Legare and P. A. Herrmann (2013), 'Cognitive consequences and constraints on reasoning about ritual', *Religion, Brain and Behavior*, 3, 63–5.
52. A. Phillips, H. M. Wellman and E. S. Spelke (2002), 'Infants' ability to connect gaze and emotional expression to intentional action', *Cognition*, 85, 53–78.
53. S. Itakura, H. Ishida, T. Kanda, Y. Shimada, H. Ishiguro et al. (2008), 'How to build an intentional android: Infant imitation of a robot's goal-directed actions', *Infancy*, 13, 519–32.
54. R. W. Byrne and A. Whiten (eds) (1988), 'Machiavellian Intelligence. Social Expertise and the Evolution of Intellect' in *Monkeys, Apes, and Humans*, Oxford: Oxford University Press.
55. A. Gopnik and J. W. Astington (1988), 'Children's understanding of representational change and its relation to the understanding of false belief and the appearance reality distinction', *Child Development*, 59, 26–37.
56. Jean Piaget and Barbel Inhelder (1956), *The Child's Conception of Space*, London: Routledge & Keegan Paul.
57. Hans Wimmer and Josef Perner (1983), 'Beliefs about beliefs: Representations and constraining function of wrong beliefs in young children's understanding of deception', *Cognition*, 13, 103–28.
58. Kristine H. Onishi and Renée Baillargeon (2005), 'Do 15-month-old infants understand false beliefs', *Science*, 308, 255–8.
59. Carla Krachun, Malinda Carpenter, Josep Call and Michael Tomasello (2009), 'A competitive nonverbal false belief task for children and apes', *Developmental Science*, 12, 521–35.

60. Susan A. Birch and Paul Bloom (2007), 'The curse of knowledge in reasoning about false beliefs', *Psychological Science*, 18, 382–6.
61. Ian Apperly (in press), 'Can theory of mind grow up? Mindreading in adults, and its implications for the development and neuroscience of mindreading' in S. Baron-Cohen, H. Tager-Flusberg and M. Lombardo (eds), *Understanding Other Minds* (third edition).
62. Lawrence A. Hirschfeld (2013), 'The Myth of Mentalizing and the Primacy of Folk Sociology' in M. R. Banaji and S. A. Gelman (eds), *Navigating the Social World*, New York: Oxford University Press.
63. David Liu and Kimberly E. Vanderbilt (2013), 'Children Learn From and About Variability Between People' in M. R. Banaji and S. A. Gelman (eds), *Navigating the Social World. What Infants, Children, and Other Species Can Teach Us*, New York: Oxford University Press.
64. C. H. Legare (2012), 'Exploring explanation: Explaining inconsistent information guides hypothesis-testing behavior in young children', *Child Development*, 83, 173–85.

CHAPTER 3

1. The exact cause of Joseph Merrick's condition is still unresolved but candidate diseases of Proteus Syndrome and neurofibromatosis Type I have been suggested.
2. M. Howell and P. Ford (1992) [1980], *The True History of the Elephant Man* (third edition), London: Penguin.
3. I. Stevenson (1992), 'A new look at maternal impressions: an analysis of 50 published cases and reports of two recent examples', *Journal of Scientific Exploration*, 6, 353–373.
4. Clarence Maloney (1976), *The Evil Eye*, New York: Columbia University Press.
5. E. A. Kensinger and D. L. Schacter (2005), 'Emotional content and reality monitoring ability: fMRI evidence for the influence of encoding processes', *Neuropsychologica*, 43, 1429–43.
6. M. Joëls, Z. Pu, O. Wiegert, M. S. Oitzl and H. J. Krugers (2006), 'Learning under stress: how does it work?', *Trends in Cognitive Science*, 10, 152–8.
7. R. Rachel Yehuda, Stephanie Mulherin Engel, Sarah R. Brand, Jonathan Seckl, Sue M. Marcus, and Gertrud S. Berkowitz (2005), 'Transgenerational effects of posttraumatic stress disorder in babies of mothers exposed to the World

Trade Center attacks during pregnancy', *Journal of Clinical Endocrinology & Metabolism*, 90, 4115–18.

8. Quote from *Discover* Magazine article published on-line 14 October 2010: http://discovermagazine.com/2010/oct/11-how-did-9-11-affect-pregnant-mothers-children
9. J. Kagan, (1994), *Galen's Prophecy: Temperament in Human Nature*, New York: Basic Books.
10. K. J. Saudino (2005), 'Behavioral genetics and child temperament', *Journal of Developmental Behavioral Pediatrics*, 26, 214–33.
11. N. A. Fox, H. A. Henderson, K. H. Rubin, S. D. Calkins and L. A. Schmidt (2001), 'Continuity and discontinuity of behavioural inhibition and exuberance: Psychophysiological and behavioural influences across the first four years of life', *Child Development*, 72, 1–21.
12. J. Bowlby (1969), *Attachment, Attachment and Loss*, Vol. 1, London: Hogarth Press.
13. K. Lorenz (1943), 'Die Angebornen Formen mogicher Erfahrung', *Zeitschrift fur Tierpsychologie*, 5, 233–409.
14. M. H. Johnson & J. Morton (1991), *Biology and Cognitive Development: The Case of Face Recognition*, Oxford: Blackwell.
15. P. S. Zeskind and B. M. Lester, (2001), 'Analysis of infant crying' in L. T. Singer and P. S. Zeskind (eds), *Biobehavioral Assessment of the Infant*, New York: Guilford, pp. 149–66.
16. E. E. Maccoby (1980), *Social Development: Psychological growth and the parent-child relationship*, New York: Harcourt Brace Janovich.
17. H. F. Harlow (1958), 'The nature of love', *American Psychologist*, 13, 573–685.
18. M. Rutter, T. G. O'Connor and The English and Romanian Adoptees (ERA) Study Team (2004), 'Are there biological programming effects for psychological development? Findings from a study of Romanian adoptees', *Developmental Psychology*, 40, 81–94.
19. J. A. Whitson and A. D. Galinsky (2008), 'Lacking control increases illusory pattern perception', *Science*, 322, 115–17.
20. T. V. Salomons, T. Johnstone, M. Backonja and R. J. Davidson (2004), 'Perceived controllability modulates the neural response to pain', *Journal of Neuroscience*, 24, 7199–203.
21. L. Murray, A. Fiori-Cowley, R. Hooper and P. Cooper (1996), 'The impact of postnatal depression and associated adversity on early mother-infant interactions and later infant outcome', *Child Development*, 67, 2512–26.
22. C. M. Pariante and A. H. Miller (2001), 'Glucocorticoid receptors in major depression: Relevance to pathophysiology and treatment', *Biological Psychiatry*, 49, 391–404.

23. C. S. de Kloet, E. Vermetten, E. Geuze, A. Kavelaars, C. J. Heijnen and H. G. M. Westenberg (2006), 'Assessment of HPA-axis function in posttraumatic stress disorder: pharmacological and non-pharmacological challenge tests, a review', *Journal of Psychiatric Research*, 40, 550–67.
24. A. K. Pesonen, K. Räikkönen, K. Feldt, K. Heinonen, C. Osmond, D. I. Phillips, D. J. Barker, J. G. Eriksson and E. Kajantie (2010), 'Childhood separation experience predicts HPA axis hormonal responses in late adulthood: a natural experiment of World War II', *Psychoneuroendocrinology*, 35, 758–67.
25. S. Clarke, D. J. Wittwer, D. H. Abbott and M. L. Schneider (1994), 'Long-term effects of prenatal stress on HPA axis activity in juvenile rhesus monkeys', *Developmental Neurobiology*, 27, 257–69.
26. Alice Graham, Phil Fisher and Jennifer Pfeifer (2013), 'What sleeping babies hear: A functional MRI study of interparental conflict and infants' emotion processing', *Psychological Science*, 24, 782–9.
27. L. Trut, I. Oskina and A. Kharlamova (2009), 'Animal evolution during domestication: the domesticated fox as a model', *BioEssays*, 31, 349–60.
28. Molly Crockett (2009), 'Values, Empathy, and Fairness across Social Barriers', *Annals of the New York Academy of Sciences*, 1167, 76–86.
29. L. N. Trut, I. Z. Plyusnina and I. N. Oskina (2004), 'An experiment on fox domestication and debatable issues of evolution of the dog', *Russian Journal of Genetics*, 40, 644–55.
30. This is known as the James–Lange theory after Carl Lange developed William James's initial proposal.

    C. G. Lange and W. James (1922), *The Emotions*, Baltimore, MD: Williams & Wilkins.
31. This alternative to the James–Lange theory was the Cannon–Bard theory after W. B. Cannon (1929), 'The James–Lange theory of emotion: A critical examination and alternative theory', *American Journal of Psychology*, 39, 106–24.

    P. Bard (1934), 'On emotional experience after decortication with some remarks on theoretical views', *Psychological Review*, 41, 309–29.
32. Joseph LeDoux (1998), *The Emotional Brain*, London: Weidenfeld & Nicolson.
33. S. Schacter and J. E. Singer (1962), 'Cognitive, social and psychological determinants of emotional state', *Psychological Review*, 69, 379–99.
34. Ian Penton-Voak, Jamie Thomas, Suzanne Gage, Mary McMurran, Sarah McDonald and Marcus Munafo (2013), 'Increasing recognition of happiness in ambiguous facial expressions reduces anger and aggressive behavior', *Psychological Science*, 24, 688–97.

35. R. M. Sullivan, M. Landers, B. Yeaman and D. A. Wilson (2000), 'Good memories of bad events in infancy: Ontogeny of conditioned fear and the amygdala', *Nature*, 407, 38–9.
36. S. Moriceau and R. M. Sullivan (2006), 'Maternal presence serves as a switch between learning fear and attraction in infancy', *Nature Neuroscience*, 9, 1004–6.
37. Sarah L. Master, Naomi I. Eisenberger, Shelley E. Taylor, Bruce D. Naiboff, David Shirinyan and Matthew D. Leiberman (2009), 'A picture's worth: Partner photographs reduce experimentally induced pain', *Psychological Science*, 20, 1316–18.
38. Dean Jensen (2006), *The Lives and Loves of Daisy and Violet Hilton: A True Story of Conjoined Twins*, Berkeley, CA: Ten Speed Press.
39. Judith Rich Harris (2006), *No Two Alike: Human Nature and Human Individuality*, W. W. Norton.
40. Guttal et al. (2012), 'Cannibalism can drive the evolution of behavioural phase polyphenism in locusts', *Ecology Letters*, 15, 1158–66.
41. David Sheldon Cohen, A. J. Tyrrell and Andrew P. Smith (1991), 'Psychological stress and susceptibility to the common cold', *New England Journal of Medicine*, 325, 606–12.
42. S. W. Cole, L. C. Hawkley, J. M. Arevalo, C. Y. Sung, R. M. Rose and J. T. Cacioppo (2007), 'Social regulation of gene expression in human leukocytes', *Genome Biology*, 8, R189.
43. Steve W. Cole (2009), 'Social regulation of human gene expression', *Current Directions in Psychological Science*, 18, 132–7.
44. R. Simmons and R. Altwegg (2010), 'Necks-for-sex or competing browsers? A critique of ideas on the evolution of the giraffe', *Journal of Zoology*, 282, 6–12.
45. F. A. Champagne, D. D. Francis, A. Mar and M. J. Meaney (2003), 'Naturally-occurring variations in maternal care in the rat as a mediating influence for the effects of environment on the development of individual differences in stress reactivity', *Physiology & Behavior*, 79, 359–71.
46. D. D. Francis, J. Diorio, D. Liu and M. J. Meaney (1999), 'Nongenomic transmission across generations in maternal behavior and stress responses in the rat', *Science*, 286, 1155–8.
47. M. J. Meaney (2001), 'The development of individual differences in behavioral and endocrine responses to stress', *Annual Review of Neuroscience*, 24, 1161–92.
48. P. O. McGowan, M. Suderman, A. Sasaki, T. C. Huang, M. Hallett, M. J. Meaney et al. (2011), 'Broad epigenetic signature of maternal care in the brain of adult rats', PLoS ONE, 6, e14739.

49. P. O. McGowan, A. Sasaki, A. C. D'Alessio, S. Dymov, B. Labonte, M. Szyf, G. Turecki and M. J. Meaney (2009), 'Epigenetic regulation of the glucocorticoid receptor in human brain associates with childhood abuse', *Nature Neuroscience* 12, 342–8.
50. Marilyn J. Essex, W. Tom Boyce, Clyde Hertzman, Lucia L. Lam, Jeffrey M. Armstrong, Sarah M. Neumann and Michael S. Kobor (2013), 'Epigenetic Vestiges of early developmental adversity: Childhood stress exposure and DNA methylation in adolescence', *Child Development*, 84, 58–75.
51. H. G. Brunner, M. Nelen, X. O. Breakefield, H. H. Ropers and B. A. van Oost (1993), 'Abnormal behavior associated with a point mutation in the structural gene for monoamine oxidase A', *Science*, 262, 578–80.
52. Ann Gibbons (2004), 'Tracking the evolutionary history of a "warrior" gene', *Science*, 304, 5672.
53. R. A. Lea, D. Hall, M. Green and C. K. Chambers, 'Tracking the evolutionary history of the warrior gene in the South Pacific', presented at the Molecular Biology and Evolution Conference in Auckland, June 2005, and the International Congress of Human Genetics, Brisbane, August 2006.
54. Rose McDermott, Dustin Tingley, Jonathan Cowden, Giovanni Frazzetto and Dominic D. P. Johnson (2009), 'Monoamine oxidase A gene (MAOA) predicts behavioral aggression following provocation', Proceedings of the National Academy. www.pnas.org_cgi_doi_10.1073_pnas.0808376106
55. Ed Yong (2010), 'Dangerous DNA: The truth about the "warrior gene"', *New Scientist*, 7 April 2010.
56. A. Caspi, J. McClay, T. E. Moffitt, J. Mill, J. Martin, I. W. Craig, A. Taylor, and R. Poulton (2002), 'Role of genotype in the cycle of violence in maltreated children', *Science*, 297, 851–4.

CHAPTER 4

1. E. Macphail (1982), *Brain and Intelligence in Vertebrates*, Oxford, England: Clarendon Press.
2. R. A. Barton and C. Venditti (2013), 'Human frontal lobes are not relatively large', Proc Natl Acad Sci USA 110, 9001–6.
3. Jeffrey Rogers et al. (2010), 'On the genetic architecture of cortical folding and brain volume in primates', *NeuroImage*, 53, 1103–8.
4. Kate Teffer and Katerina Semendeferi (2012), 'Human prefrontal cortex: Evolution, development, and pathology' in M. A. Hofman and D. Falk (eds), *Progress in Brain Research*, vol. 195, Elsevier.

5. J. Hill, T. Inder, J. Neil, D. Dierker, J. Harwell and D. Van Essen (2010), 'Similar patterns of cortical expansion during human development and evolution', *Proceedings of the National Academy of Sciences of the United States of America*, 107, 13135–40.
6. Xiling Liu, Mehmet Somel, Lin Tang et al. (2012), 'Extension of cortical synaptic development distinguishes humans from chimpanzees and macaques', Genome Research published online 2 February 2012: doi:10.1101/gr.127324.111
7. Robert W. Thatcher (1992), 'Cyclic cortical reorganization during early childhood', *Brain and Cognition*, 20, 24–50.
8. G. Kochanska, K. C. Coy, and K. T. Murray (2001), 'The development of self-regulation in the first four years of life', *Child Development*, 72, 1091–111.
9. Dan Gilbert (2007), *Stumbling Upon Happiness*, Perennial.
10. W. A. Roberts (2002), 'Are animals stuck in time?', *Psychological Bulletin*, 128, 473–89.
11. N. J. Mulcahy and J. Call (2010), 'Apes save tools for future use', *Science*, 312, 1038–9.
12. T. Suddendorf and J. Busby (2003), 'Mental time travel in animals?', *Trends in Cognitive Sciences*, 7, 391–5.
13. T. Suddendorf and J. Busby (2005), 'Making decisions with the future in mind: Developmental and comparative identification of mental time travel', *Learning and Motivation*, 36, 110–25.
14. Christopher M. Filley (2010), 'The frontal lobes' in Michael J. Aminoff, François Boller and Dick F. Swaab (eds), *History of Neurology*, Elsevier B.V., pp. 557–70.
15. D. O. Hebb (1977), 'Wilder Penfield: his legacy to neurology. The frontal lobe', *Canadian Medical Association Journal*, 116, 1373–4.
16. Philip David Zelazo and Stephanie M. Carlson (2012), 'Hot and cool executive function in childhood and adolescence: Development and plasticity', *Child Development Perspectives*, 6, 354–60.
17. Yuko Munakata, Seth A. Herd, Christopher H. Chatham, Brendan E. Depue, Marie T. Banich and Randall C. O'Reilly (2011), 'A unified framework for inhibitory control', *Trends in Cognitive Sciences*, 15, 453–9.
18. A. D. Smith, I. D. Gilchrist and B. M. Hood (2005), 'Children's search behaviour in large-scale space: Developmental components of exploration', *Perception*, 34, 1221–9.
19. Brenda Milner (1963), 'Effect of Different Brain Lesions on Card Sorting', *Archives of Neurology*, 9, 90–100.
20. Adele Diamond (1991), 'Neuropsychological insights into the meaning of object concept development' in S.Carey and R. Gelman (eds), *The Epigenesis of Mind: Essays on Biology and Cognition*, Hillsdale, NJ: Lawrence Erlbaum, pp. 67–110.

21. John Ridley Stroop (1935), 'Studies of interference in serial verbal reactions', *Journal of Experimental Psychology*, 18, 643–62.
22. N. Raz, (2000), 'Aging of the brain and its impact on cognitive performance: Integration of structural and functional findings' in F. I. Criak and T. A. Salthouse (eds), *Handbook of Aging and Cognition*, Mahwah, NJ: Erlbaum, pp. 1–90.
23. P. W. Burgess and R. L. Wood (1990), 'Neuropsychology of behaviour disorders following brain injury' in R. L. Wood (ed.), *Neurobehavioural sequelae of traumatic brain injury*, New York: Taylor and Francis, pp. 110–33.
24. M. Macmillan (2000), *An Odd Kind of Fame: Stories of Phineas Gage*, Cambridge, MA: MIT Press.
25. Alexander's story is reported here: http://www.dailymail.co.uk/health/article-393938/The-freak-accident-left-son-obsessed-sex.html
    Here is the video of the marathon incident: http://www.youtube.com/watch?v=ELsGvt4Lsjo
26. S. J. Blakemore (2012), 'Imaging brain development: The adolescent brain', *Neuroimage*, 61, 397–406.
27. J. A. Fugelsang and K. N. Dunbar (2005), 'Brain-based mechanisms underlying complex causal thinking', *Neuropsychologia*, 43, 1204–13.
28. Terrie Moffitt et al. (2011), 'A gradient of childhood self-control predicts health, wealth, and public safety', *Proceedings of the National Academy of Science*, 108, 2693–8.
29. K. Meiers (2002), *Problem Schulfähigkeit. Grundschule* 5, 10–12
30. Walter Mischel, Ebbe B. Ebbesen and Antonette Raskoff Zeiss (1972), 'Cognitive and attentional mechanisms in delay of gratification', *Journal of Personality and Social Psychology*, 21, 204–18.
31. Walter Mischel, Yuichi Shoda and Monica L. Rodriguez (1989), 'Delay of gratification in children', *Science*, 244, 933–8.
32. Erik Erikson (1963), *Childhood and Society*, New York: Norton, p. 262.
33. Susan Crockenberg and Cindy Litman (1990), 'Autonomy as competence in 2-year-olds: Maternal correlates of child defiance, compliance, and self-assertion', *Developmental Psychology*, 26, 961–971.
34. Lisa Cameron, N. Erkal, L. Gangdharan and X. Meng (2013), 'Little Emperors: Behavioral impacts of China's one-child policy', *Science*, 339, 953–7.
35. Celeste Kidd, Holly Palmeri and Richard N. Aslin (2013), 'Rational snacking: Young children's decision-making on the marshmallow task is moderated by beliefs about environmental reliability', *Cognition*, 126, 109–14.

36. L. Michaelson, A. dela Vega, C. H. Chatham and Y. Munakata (2013), 'Delaying gratification depends on social trust', *Frontiers in Psychology*, 4:355. doi: 10.3389/fpsyg.2013.00355
37. I. J. Toner, L. P. Moore and B. A. Emmons (1980), 'The effect of being labeled on subsequent self-control in children', *Child Development*, 51, 618–21.
38. Richard H. Thaler and Cass R. Sunstein (2009, updated edition), *Nudge: Improving Decisions About Health, Wealth, and Happiness*, New York: Penguin.
39. L. A. Liikkanen (2008), 'Music in every mind: Commonality of involuntary musical imagery' in: K. Miyazaki, Y. Hiraga, M. Adachi, Y. Nakajima and M. Tsuzaki (eds), *Proceedings of the 10th international conference on music perception and cognition (ICMPC10)*, 408–412. Sapporo, Japan.
40. D. M. Wegner, D. J Schneider, S. R. Carter and T. L. White (1987), 'Paradoxical effects of thought suppression', *Journal of Personality and Social Psychology*, 53, 5–13.
41. C. Neil Macrae, Galen V. Bodenhausen, Alan B. Milne, Jolanda Jetten (1994), 'Out of mind but back in sight: Stereotypes on the rebound', *Journal of Personality and Social Psychology*, 67, 808–17.
42. Roger L. Albin and Jonathan W. Mink (2006), 'Recent advances in Tourette Syndrome research', *Trends in Neurosciences*, 39, 175–82.
43. I have Tourette's but Tourette's doesn't have me (2005). http://www.imdb.com/title/tt0756661/quotes
44. I. Osborn (1998), *Tormenting Thoughts and Secret Rituals: The Hidden Epidemic of Obsessive-Compulsive Disorder* (New York, NY: Dell).
45. A. M. Graybiel and S. L. Rauch (2000), 'Toward a neurobiology of obsessive compulsive disorder', *Neuron*, 28, 343–7.
46. Roy F. Baumeister and John Tierney (2011), *Willpower: Why self-control is the secret to success*, Penguin.
47. R. F. Baumeister, E. Bratslavsky, M. Muraven and D. M. Tice (1998), 'Self- control depletion: Is the active self a limited resource?', *Journal of Personality and Social Psychology*, 74, 1252–65.
48. Wilhelm Hofmann, Roy F. Baumeister, Georg Förster and Kathleen D. Vohs (2012), 'Everyday temptations: An experience sampling study of desire, conflict, and self-control', *Journal of Personality and Social Psychology*, 102, 1318–35.

CHAPTER 5

1. William Golding (1954), *Lord of the Flies*, London: Faber & Faber.
2. Taken from Golding's acceptance speech for the Nobel Prize in Literature in 1983.
3. Laura Manuel (2006), 'Relationship of personal authoritarianism with parenting styles', *Psychological Reports*, 98, 193–8.
4. Steve Pinker (2012), *The Better Angels of Our Nature: A History of Violence and Humanity*, London: Penguin.
5. http://rendezvous.blogs.nytimes.com/2012/09/02/has-the-burqa-ban-worked-in-france/ *International Herald Tribune* article, September 2012, retrieved March 2013.
6. Marc D. Hauser (2006), *Moral Minds: How Nature Designed Our Universal Sense of Right and Wrong*, New York: Harper Collins.
7. Valerie Kuhlmeier, Karen Wynn and Paul Bloom (2003), 'Attribution of dispositional states by 12-month-olds', *Psychological Science*, 14, 402–8.
8. J. Kiley Hamlin, Karen Wynn and Paul Bloom (2007), 'Social evaluation by preverbal infants', *Nature*, 450, 557–9.
9. J. Kiley Hamlin, Karen Wynn, Paul Bloom and Neha Mahajan (2011), 'How infants and toddlers react to antisocial others', *Proceedings of the National Academy*, 108, 19931–6.
10. K.A. Dunfield and V.A. Khulmeier (2010), 'Intention-mediated selective helping in infancy', *Psychological Science*, 21, 523–7.
11. V. A. Khulmeier (2013), 'Disposition attribution in infancy' in M. R. Banaji and S. A. Gelman (eds), *Navigating the Social World. What Infants, Children, and Other Species Can Teach Us*, New York: Oxford University Press.
12. Paul Bloom (2013), *Just Babies: The Origins of Good and Evil*. New York: Crown.
13. C. U. Shantz (1987), 'Conflicts between children', *Child Development*, 58, 283–305.
14. D. F. Hay and H. S. Ross (1982), 'The social nature of early conflict', *Child Development*, 53, 105–13.
15. 2013: http://www.thesun.co.uk/sol/homepage/news/4977445/Man-dragged-50ft-along-road-after-trying-to-stop-car-theft.html

    2012: http://www.newsnet5.com/dpp/news/local_news/oh_cuyahoga/car-thieves-crash-stolen-car-killing-owner-who-was-hanging-onto-the-hood

    http://www.tulsaworld.com/article.aspx/Woman_who_died_trying_to_prevent_car_theft_remembered/20120310_11_a1_cutlin972357

16. William James (1890), *Principles of Psychology*, New York: Henry Holt, p. 291.
17. R. Belk (1988), 'Possessions and the extended self', *Journal of Consumer Research*, 15, 139–68.
18. J. E. Stake (2004), 'The property "instinct"', *Philosophical Transactions of the Royal Society B: Biological Sciences*, 359, 1763–74.
19. M. Rodgon and S. Rashman (1976), 'Expression of owner-owned relationships among holophrastic 14- to 32-month-old children', *Child Development*, 47, 1219–22.
20. B. Hood and P. Bloom (2008), 'Children prefer certain individuals over perfect duplicates', *Cognition*, 106, 455–62.
21. H. Ross, C. Conant and M. Vickar (2011), Property rights and the resolution of social conflict, *New Directions for Child and Adolescent Development*, 132, 53–64.
22. F. Rossano, H. Rakoczy and M Tomasello (2011), 'Young children's understanding of violations of property rights', *Cognition*, 121, 219–27.
23. O. Friedman, J. W. van de Vondervoort, M. A. Defeyter and K. R. Neary (2013), 'First possession, history, and young children's ownership judgments', *Child Development*, 84, 1519–25.
24. http://www.cbc.ca/news/yourcommunity/2013/02/bansky-graffiti-ripped-off-london-wall-put-on-auction-in-us.html
25. P. Kanngiesser, N. L. Gjersoe and B. M. Hood (2010), 'Transfer of property ownership following creative labour in preschool children and adults', *Psychological Science*, 21, 1236–41.
26. K. R. Olson and A. Shaw (2011), '"No fair, copycat!": what children's response to plagiarism tells us about their understanding of ideas', *Developmental Science*, 14, 431–9.
27. D. J. Turk, K. van Bussel, G. D. Waiter and C. N. Macrae (2011), 'Mine and me: Exploring the neural basis of object ownership', *Journal of Cognitive Neuroscience*, 23, 3657–68.
28. S. J. Cunningham, D. J Turk and C. N. Macrae (2008), 'Yours or mine? Ownership and memory', *Consciousness and Cognition*, 17, 312–18.
29. R. Thaler (1980), 'Toward a positive theory of consumer choice', *Journal of Economic Behavior and Organization*, 1, 39–60.
30. J. Heyman, Y. Orhun and D. Ariely (2004), 'Auction fever: the effect of opponents and quasi-endowment on product valuations', *Journal of Interactive Marketing*, 18, 7–21.
31. J. R. Wolf, H. R. Arkes and W. A. Muhanna (2008), 'The power of touch: An examination of the effect of duration of physical contact on the valuation of objects', *Judgement and Decision Making*, 3, 476–82.

32. D. Kahneman, J. L. Knetsch and R. H. Thaler (1991), 'Anomalies: The endowment effect, loss aversion and status quo bias', *Journal of Economic Perspectives*, 5, 193–206.
33. B. Knutson, G. E. Wimmer, S. Rick, N. G. Hollon, D. Prelec and G. Loewenstein (2008), 'Neural antecedents of the endowment effect', *Neuron*, 58, 814–22.
34. W. T. Harbaugh, K. Krause and L. Vesterlund (2001), 'Are adults better behaved than children? Age, experience, and the endowment effect', *Economics Letters*, 70, 175–81.
35. M. Wallendorf and E. J. Arnould (1988), '"My favourite things": A cross-cultural inquiry into object attachment, possessiveness, and social linkage', *Journal of Consumer Research*, 14, 531–47.
36. Coren L. Apicella, Eduardo M. Azevedo, James H. Fowler and Nicholas A. Christakis (2013), 'Evolutionary Origins of the Endowment Effect: Evidence from Hunter-Gatherers', *American Economic Review*, 23 August 2013. Available at SSRN: http://ssrn.com/abstract=2255650 or http://dx.doi.org/10.2139/ssrn.2255650.
37. L. L. Birch and J. Billman (1986), 'Preschool children's food sharing with friends and acquaintances', *Child Development*, 57, 387–95.
38. M. Gummerum, Y. Hanoch, M. Keller, K. Parsons and A. Hummel (2010), 'Preschoolers' allocations in the dictator game: The role of moral emotions', *Journal of Economic Psychology*, 31, 25–34.
39. Ernst Fehr, Helen Bernhard and Bettina Rockenbach (2008), 'Egalitarianism in young children', *Nature*, 454, 1079–1084.
40. Katharina Hamann, Felix Warneken, Julia R. Greenberg and Michael Tomasello (2012), 'Collaboration encourages equal sharing in children but not in chimpanzees', *Nature*, 476, 328–31.
41. P. Blake and D. Rand (2010), 'Currency value moderates equity preference among young children', *Evolution and Human Behavior*, 31, 210–18.
42. David Reinstein and Gerhard Riener (2012), 'Reputation and influence in charitable giving: an experiment', *Theory and Decision*, 72, 221–43.
43. F. Alpizar, F. Carlsson and O. Johansson-Stenman (2008), 'Anonymity, reciprocity, and conformity: Evidence from voluntary contributions to a national park', *Journal of Public Economics*, 92, 1047–1060.
44. K. L. Leimgruber, A. Shaw, L. R. Santos and K. R. Olson (2012), 'Young Children Are More Generous When Others Are Aware of Their Actions', *PLoS ONE* 7(10): e48292. doi:10.1371/journal.pone.0048292
45. Felix Warneken and Michael Tomasello (2006), 'Altruistic helping in human infants and young chimpanzees', *Science*, 311, 1301–2.

46. M R. Lepper, D. Greene and R. E. Nisbett (1973), 'Undermining children's intrinsic interest with an extrinsic reward: A test of the "overjustification" hypothesis', *Journal of Personality and Social Psychology*, 28, 129–37.
47. F. Warneken (2013), 'What do children and chimpanzees reveal about human altruism?' in M. R. Banaji and S. A. Gelman (eds), *Navigating the Social World. What Infants, Children, and Other Species Can Teach Us*, New York: Oxford University Press.
48. Joan Silk, Commentary on Mike Tomasello (2009), *Why We Cooperate*, Boston: Boston Review.
49. F. Warneken, B. Hare, A. P. Melis, D. Haus and M. Tomasello, (2007), 'Spontaneous altruism by chimpanzees and young children', *PLoS Biology*, 5, 1414–20.
50. Judith M. Burkart, Ernst Fehr, Charles Efferson and Carel P. van Schaik (2007), 'Other-regarding preferences in a non-human primate: Common marmosets provision food altruistically', *Proceedings of the National Academy*, 104, 19762–6.
51. A. Ueno and T. Matsuzawa (2004), 'Food transfer between chimpanzee mothers and their infants', *Primates*, 45, 231–9.
52. William T. Harbaugh, Ulrich Mayr and Daniel R. Burghart (2007), 'Neural responses to taxation and voluntary giving reveal motives for charitable donations', *Science*, 316, 1622–5.
53. Ernst Fehr and Simon Gächter (2002), 'Altruistic punishment in humans', *Nature*, 415, 137–40.
54. W. Guth, R. Schmittberger and B. Schwarze (1982), 'An experimental analysis of ultimatum bargaining', *Journal of Economic Behavior & Organization*, 3, 367–88.
55. A. G. Sanfey, J. K. Rilling, J. A. Aronson, L. E. Nystrom and J. D. Cohen (2003), 'The neural basis of economic decision-making in the Ultimatum Game', *Science*, 300, 1755–8.
56. D. Knoch, A. Pascual-Leone, K. Meyer, V. Treyer and E. Fehr (2006), 'Diminishing reciprocal fairness by disrupting the right prefrontal cortex', *Science*, 314, 829–32.
57. K. Jensen, J. Call and M. Tomasello (2007), 'Chimpanzees are maximizers in an ultimatum game', *Science*, 318, 107–9.
58. Sarah F. Brosnan and Frans de Waal (2003), 'Monkeys reject unequal pay', *Nature*, 425, 297–9.
59. J. Bräuer, J. Call and M. Tomasello (2006), 'Are apes really inequity averse?', *Proceedings of the Royal Society B*, 273, 3123–8.
60. Dan Ariely (2008), *Predictably Irrational*, New York: HarperCollins.

61. John Nash (1951), 'Non-cooperative Games', *Annals of Mathematics*, 54, 286–95.
62. Richard Dawkins (1976), *The Selfish Gene*, Oxford University Press.
63. C. Adami and A. Hintze, 'Evolutionary instability of zero determinant strategies demonstrates that winning is not everything', *Nature Communications*, 4, 2193. doi: 10.1038/ncomms3193 (2013).
64. Paul Slovic (2007), '"If I look at the mass I will never act": Psychic numbing and genocide', *Judgement and Decision Making*, 2, 79–95.
65. Karen E. Jenni and George Loewenstein (1997), 'Explaining the "Identifiable victim effect"', *Journal of Risk and Uncertainty*, 14, 235–57.
66. D. Västfjäll, E. Peters and P. Slovic (in preparation), 'Representation, affect, and willingness-to-donate to children in need', unpublished manuscript.
67. Leon R. Kass (1997), 'The Wisdom of Repugnance', *The New Republic*, 216, 17–26.
68. Jesse Bering (2013), *Perv: The Sexual Deviant in All of Us*, Scientific American/Farrar, Straus and Giroux.
69. Jonathan Haidt (2001), 'The emotional dog and its rational tail: A social intuitionist approach to moral judgement', *Psychological Review*, 108: 814–34.
70. J. Thomson (1985), 'The Trolley Problem', *Yale Law Journal*, 94, 1395–1415.
71. J. D. Greene, R. B. Sommerville, L. E. Nystrom, J. M. Darley and J. D. Cohen (2001), 'An fMRI investigation of emotional engagement in moral judgement', *Science*, 293, 2105–8.
72. William B. Swann, Jr., Ángel Gómez, John F. Dovidio, Sonia Hart and Jolanda Jetten (2010), 'Dying and killing for one's group: Identity fusion moderates responses to intergroup versions of the trolley problem', *Psychological Science*, 21, 1176–83.
73. Joshua Greene (2007), 'The secret joke of Kant's soul', in W. Sinnott-Armstrong (ed.), *Moral Psychology, Vol. 3: The Neuroscience of Morality: Emotion, Disease, and Development*, Cambridge, MA: MIT Press.
74. Jean Piaget (1932/1965). *The Moral Judgement of the Child*, New York: Free Press.
75. Lawrence Kohlberg (1963), 'Development of children's orientation towards a moral order (Part I). Sequencing in the development of moral thought', *Vita Humana*, 6, 11–36.
76. B. M. DePaulo and D. A. Kashy (1998), 'Everyday lies in close and casual relationships', *Journal of Personality and Social Psychology*, 74, 63–79.
77. B. M. DePaulo, D. A. Kashy, S. E. Kirkendol, M. M. Wyer and J. A. Epstein (1996), 'Lying in everyday life', *Journal of Personality and Social Psychology*, 70, 979–95.

78. William von Hippel and Robert Trivers (2011), 'The evolution and psychology of self-deception', *Behavioral and Brain Sciences*, 34, 1–56.
79. Robert Trivers (1976), Foreword, in R. Dawkins, *The Selfish Gene*, Oxford University Press, pp. 19–20.
80. M. D. Alicke and C. Sedikides (2009), 'Self-enhancement and self-protection: What they are and what they do', *European Review of Social Psychology*, 20, 1–48.
81. Tali Sharot (2012), *The Optimism Bias: A Tour of the Irrationally Positive Brain*, London: Vintage.
82. C. Ward Struthers, Judy Eaton, Alexander G. Santelli, Melissa Uchiyama and Nicole Shirvani (2008), 'The effects of attributions of intent and apology on forgiveness: When saying sorry may not help the story', *Journal of Experimental Social Psychology*, 44, 983–92.
83. S. Harris, S. A. Sheth and M. S. Cohen (2007), 'Functional Neuroimaging of belief, disbelief and uncertainty', *Annals of Neurology*, 63, 141–7.
84. M. Main and C. George (1985), 'Responses of young abused and disadvantaged toddlers to distress in age mates', *Developmental Psychology*, 21, 407–12.
85. S. Johnson, C. S. Dweck and F. Chen (2007), 'Evidence for infants' internal working models of attachment', *Psychological Science*, 18, 501–2.

CHAPTER 6

1. Shane Bauer, 'Solitary in Iran nearly broke me. Then I went inside America's prisons', http://www.motherjones.com/politics/2012/10/solitary-confinement-shane-bauer, *Mother Jones*, December 2012, retrieved October 2013.
2. Nelson Mandela (1994), *Long Walk to Freedom*, London: Little Brown, p. 52.
3. Reuters, 'U.S. Bureau of Prisons to review solitary confinement', http://www.nytimes.com/reuters/2013/02/04/us/04reuters-usa-prisons-solitary.html?ref=solitaryconfinement, *New York Times*, February 2013, retrieved February 2013.
4. Joshua Foer and Michel Siffre (2008), 'Caveman: An Interview with Michel Siffre', http://www.cabinetmagazine.org/issues/30/foer.php, *Cabinet Magazine*, issue 30.
5. Michel Siffre, 'Six Months Alone in a Cave,' *National Geographic* (March 1975), 426–435.
6. J. S. House, K. R. Landis and D. Umberson (1988), 'Social relationships and health', *Science*, 241, 540–45.

7. John T. Cacioppo, James H. Fowler and Nicholas A. Christakis (2009), 'Alone in the crowd: The structure and spread of loneliness in a large social network', *Journal of Personality and Social Psychology*, 97, 977–91.
8. Charles Darwin (1872), *The Expression of the Emotions in Man and Animals*, London: John Murray.
9. J. M. Susskind et al. (2008), 'Expressing fear enhances sensory acquisition', *Nature Neuroscience*, 11, 843–50.
10. V. Gallese, L. Fadiga, L. Fogassi and G. Rizzolatti (1996), 'Action recognition in the premotor cortex', *Brain*, 119, 593–609.
11. C. Keysers (2011), *The Empathic Brain*, Los Gatos, CA: Smashwords ebook.
12. This claim was made by the eminent neuroscientist Vilayanur Ramachandran and is related in C. Keysers (2011), *The Empathic Brain*, Los Gatos, CA: Smashwords ebook.
13. R. Mukamel, A. D. Ekstrom, J. Kaplan, M. Iacoboni and I. Fried (2010), 'Single-neuron responses in humans during execution and observation of actions', *Current Biology*, 20, 750–56.
14. Cecilia Heyes (2010), 'Where do mirror neurons come from?', *Neuroscience and Biobehavioral Reviews*, 34, 575–83.
15. Neha Mahajan & Karen Wynn (2012), 'Origins of "Us" versus "Them": Prelinguistic infants prefer similar others', *Cognition*, 124, 227–33.
16. Gordon Gallup (1970), 'Chimpanzees: Self-recognition', *Science*, 167, 86–7. One curious exception is the domesticated dog, for reasons that are unknown. Apparently, domestication by man to be sociable did not include this component of social awareness.
17. B. Amsterdam (1972), 'Mirror self-image reactions before age two', *Developmental Psychobiology*, 5, 297–305.
18. Charles Darwin (1872), *The Expression of the Emotions in Man and Animals.* London: John Murray, p. 325.
19. Mark R. Leary, Thomas W. Britt, William D Cutlip and Janice L. Templeton (1992), 'Social blushing', *Psychological Bulletin*, 112, 446–60.
20. K. M. Zosuls, D. Ruble, C. S. Tamis-LeMonda, P. E. Shrout, M. H. Bornstein and F. K. Greulich (2009), 'The acquisition of gender labels in infancy: Implications for sex-typed play', *Developmental Psychology*, 45, 688–701.
21. P. C. Quinn, J. Yahr , A. Kuhn, A. M. Slater and O. Pascalis (2002), 'Representation of the gender of human faces by infants: A preference for female', *Perception* 31, 1109–21.
22. E. E. Maccoby and C. N. Jacklin (1987), 'Gender segregation in childhood' in E. H. Reese (ed.), *Advances in child development and behavior*, vol. 20, New York: Academic Press, pp. 239–87.

23. H. Abel and R. Sahinkaya (1962), 'Emergence of sex and race friendship preferences', *Child Development*, 33, 939–43.
24. C. F. Miller, C. L. Martin, R. A. Fabes and L. D. Hanish (2013), 'Bringing the Cognitive and Social Together' in M. R. Banaji and S. A. Gelman (eds), *Navigating the Social World. What Infants, Children, and Other Species Can Teach Us*, New York: Oxford University Press.
25. R. S. Bigler, C. S. Brown and M. Markell, M. (2001), 'When groups are not created equal: Effects of group status on the formation of intergroup attitudes in children', *Child Development*, 72, 1151–62.
26. R. S. Bigler (2013) 'Understanding and Reducing Social Stereotyping and Prejudice Among Children' in M. R. Banaji and S. A. Gelman (eds), *Navigating the Social World. What Infants, Children, and Other Species Can Teach Us*, New York: Oxford University Press, p. 328.
27. Adam Waytz & Jason P. Mitchell (2011), 'Two mechanisms for simulating other minds: Dissociations between mirroring and self-projection', *Current Directions in Psychological Science*, 20, 197–200.
28. Xiaojing Xu, Xiangyu Zuo, Xiaoying Wang and Shihui Han (2009), 'Do you feel my pain? Racial group membership modulates empathic neural responses', *Journal of Neuroscience*, 29, 8525–9.
29. Adrianna C. Jenkins and Jason P. Mitchell (2011), 'Medial prefrontal cortex subserves diverse forms of self-reflection', *Social Neuroscience*, 6, 211–18.
30. Kyungmi Kim and Marcia K. Johnson (2012), 'Extended self: medial prefrontal activity during transient association of self and objects', *Scan*, 7, 199–207.
31. M. Stel, J. Blascovich, C. McCall, J. Mastop, R. B. Van Baaren and R. Vonk (2010), 'Mimicking disliked others: Effects of a priori liking on the mimicry-liking link', *European Journal of Social Psychology*, 40, 867–80.
32. Xiaojing Xu, Xiangyu Zuo, Xiaoying Wang and Shihui Han (2009), 'Do you feel my pain? Racial group membership modulates empathic neural responses', *Journal of Neuroscience*, 29, 8525–9.
33. Stanley Milgram (1963), 'Behavioral study of obedience', *Journal of Abnormal and Social Psychology*, 67, 371–8.
34. P. Zimbardo (2007), *The Lucifer Effect: How Good People Turn Evil*, London: Random House.
35. H. Tajfel, M. G. Billig, R. P. Bundy and C. Flament (1971), 'Social categorization and intergroup behaviour', *European Journal of Social Psychology*, 1, 149–78.
36. Martin Niemöller, 'Als die Nazis die Kommunisten holten . . .', http://www.martin-niemoeller-stiftung.de/4/daszitat/a31
37. S. Alexander Haslam and Stephen D. Reicher (2012), 'Contesting the

"Nature" of Conformity: What Milgram and Zimbardo's studies really show', PLOS Biology, volume 10, issue 11, e1001426.

38. N. Mahajan, M. A. Martinez, N. L. Gutierrez, G. Diesendruck, M. R. Banaji and L. R. Santos (2011), 'The evolution of intergroup bias: perceptions and attitudes in rhesus macaques', *Journal of Personality & Social Psychology*, 100, 387–405.
39. Kate Fox (2004), *Watching the English: The Hidden Rules of English Behaviour*, Hodder & Stoughton, London.
40. Solomon E. Asch (1956), 'Studies of independence and conformity: A minority of one against a unanimous majority', *Psychological Monographs: General and Applied*, 70, 1–70.
41. R. Bond and P. Smith (1996), 'Culture and conformity: A meta-analysis of studies using Asch's (1952b, 1956) line judgement task', *Psychological Bulletin*, 119, 111–37.
42. Jamil Zaki, Jessica Schirmer and Jason P. Mitchell (2011), 'Social influence modulates the neural computation of value', *Psychological Science*, 22, 894–900.
43. J. Cloutier, T. F. Heatherton, P. J. Whalen and W. M. Kelley (2008), 'Are attractive people rewarding? Sex differences in the neural substrates of facial attractiveness', *Journal of Cognitive Neuroscience*, 20, 941–51.
44. R.B. Cialdini (2005), 'Don't throw away the towel: Use social influence research', *American Psychological Society*, 18, 33–4.
45. Richard H. Thaler and Cass R. Sunstein (2009, updated edition), *Nudge: Improving Decisions About Health, Wealth, and Happiness*, New York: Penguin.
46. Leon Festinger (1957), *A Theory of Cognitive Dissonance*. Stanford, CA: Stanford University Press.
47. Vincent van Veen, Marie K. Krug, Jonathan W. Schooler and Cameron S. Carter (2009), 'Neural activity predicts attitude change in cognitive dissonance', *Nature Neuroscience*, 12, 1469–75.
48. Ellen J. Langer (1978), 'Rethinking the Role of Thought in Social Interaction' in John H. Harvey, William J. Ickes and Robert F. Kidd (eds), *New Directions in Attribution Research*, vol. 2, Lawrence Erlbaum Associates, pp. 35–58.
49. Quoted in Carol Tavris and Elliot Aronson (2007), *Mistakes Were Made (but not by me): Why We Justify Foolish Beliefs, Bad Decisions and Hurtful Acts*, Harcourt Inc.
50. A. G. Greenwald and M. R. Banaji (1995), 'Implicit social cognition: Attitudes, self-esteem, and stereotypes', *Psychological Review*, 102, 4–27.
51. Andreas Olsson, Jeffrey P. Ebert, Mahzarin R. Banaji and Elizabeth A. Phelps

(2005), 'The role of social groups in the persistence of learned fear', *Science*, 309, 785–7.

52. Carlos David Navarrete, Andreas Olsson, Arnold K. Ho, Wendy Berry Mendes, Lotte Thomsen and James Sidanius (2009), 'Fear extinction to an out-group face: The role of target gender', *Psychological Science*, 20, 155–8.
53. L.F. Pendry (2008), 'Social cognition' in M. R. Hewstone, W. Stroebe and K. Jonas (eds), *Introduction to Social Psychology* (fourth edition), Oxford: Blackwell, pp. 67–87.
54. Daniel Kahneman (2011), *Thinking Fast, Thinking Slow*, Farrar, Straus and Giroux.
55. J. Correll, B. Park, C. M. Judd and B. Wittenbrink (2002), 'The police officer's dilemma: Using ethnicity to disambiguate potentially threatening individuals', *Journal of Personality and Social Psychology*, 83, 1314–29.
56. M. Snyder. and W. B. Swann, Jr. (1978), 'Hypothesis testing processes in social interaction', *Journal of Personality and Social Psychology*, 36, 1202–12.
57. D. F. Halpern (2004), 'A cognitive-process taxonomy for sex differences in cognitive abilities', *Current Directions in Psychological Science*, 13, 135–9.
58. 'It's official: women are actually better parkers than men': http://www.ncp.co.uk/documents/pressrelease/ncp-parking-survey.pdf
59. S. J. Spencer, C. M. Steele and D. M. Quinn (1999), 'Stereotype threat and women's math performance', *Journal of Experimental Social Psychology*, 35, 4–28.
60. C. M. Steele and J. Aronson (1995), 'Stereotype threat and the intellectual test performance of African Americans', *Journal of Personality and Social Psychology*, 69, 797–811.
61. B. M. Hood, N. L. Gjersoe, K. Donnelly, A. Byers and S. Itajkura (2011), 'Moral contagion attitudes towards potential organ transplants in British and Japanese adults', *Journal of Culture and Cognition*, 11, 269–86.
62. C. Dyer (1999), 'English teenager given heart transplant against her will', *British Medical Journal*, 319(7204): 209.
63. M. A. Sanner (2005), 'Living with a stranger's organ: Views of the public and transplant recipients', *Annals of Transplantation*, 10, 9–12.
64. D. L. Medin and A. Ortony (1989), 'Psychological Essentialism' in S. Vosniadou and A. Ortony (eds), *Similarity and Analogical Reasoning*, Cambridge University Press.
65. Susan A. Gelman (2003), *The Essential Child: Origins of Essentialism in Everyday Thought*, Oxford University Press.
66. Susan A. Gelman and Henry M. Wellman (1991), 'Insides and essences: early understandings of the non-obvious', *Cognition*, 38, 213–44.

67. Susan A. Gelman and Ellen M. Markman (1986), 'Categories and induction in young children', *Cognition*, 23, 183–209.
68. Meredith Meyer, Susan A. Gelman and Sarah-Jane Leslie (submitted), 'My heart made me do it: Children's essentialist beliefs about heart transplants'.
69. N. Haslam, B. Bastian, P. Bain and Y. Kashima (2006), 'Psychological essentialism, implicit theories, and intergroup relations', *Group Processes and Intergroup Relations*, 9, 63–76.
70. Gil Diesendruck (2013), 'Essentialism: The development of a simple, but potentially dangerous, idea' in M. R. Banaji and S. A. Gelman (eds), *Navigating the Social World. What Infants, Children, and Other Species Can Teach Us*, New York: Oxford University Press.
71. N. I. Eisenberger, M. D. Lieberman and K. D. Williams (2003), 'Does rejection hurt? An FMRI study of social exclusion', *Science*, 302, 290–92.
72. Nicole Legate, Cody R. DeHaan, Netta Weinstein and Richard M. Ryan (2013), 'Hurting you hurts me too: The psychological costs of complying with ostracism', *Psychological Science*, 24, 583–8.
73. K. D. Williams and S. A. Nida, (2011), 'Ostracism: consequences and coping', *Current Directions in Psychology*, 20, 71–5.
74. Lowell Gaertner, Jonathan Iuzzini and Erin M. O'Mara (2008), 'When rejection by one fosters aggression against many: Multiple-victim aggression as a consequence of social rejection and perceived groupness', *Journal of Experimental Social Psychology*, 44, 958–70.
75. W. A. Warburton, K. D. Williams and D. R. Cairns (2006), 'When ostracism leads to aggression: The moderating effects of control deprivation', *Journal of Experimental Social Psychology*, 42, 213–20.
76. Centers for Disease Control and Prevention, 'Youth suicide'. http://www.cdc.gov/violenceprevention/pub/youth_suicide.html Accessed 22 October 2013.
77. A. Brunstein Klomek, F. Marrocco, M. Kleinman, I.S. Schonfeld and M.S. Gould (2007), 'Bullying, depression, and suicidality in adolescents', *Journal of the American Academy of Child Adolescent Psychiatry*, 46, 40–49. http://www.cdc.gov/violenceprevention/pub/youth_suicide.html
78. Marcel F. van der Wal, Cees A. M. de Wit, Remy A. Hirasing (2003), Psychosocial health among young victims and offenders of direct and indirect bullying,. *Pediatrics*, 111, 1312–17.
79. Julie A. Paquette and Marion K. Underwood (1999), 'Gender differences in young adolescents' experiences of peer victimization: social and physical aggression', *Merrill-Palmer Quarterly*: Vol. 45: Issue 2, Article 5.
80. M. Boulton (1997), 'Teachers' views on bullying: Definitions, attitudes and ability to cope', *British Journal of Educational Psychology*, 67, 223–33.

81. K. D. Williams (2009), 'Ostracism: A temporal need-threat model', in M. Zanna (ed.), *Advances in Experimental Social Psychology*, New York, Academic Press, 279–314.

EPILOGUE

1. Richard E. Nisbett (2003), *The Geography of Thought*, Nicholas Brealey Publishing.
2. Jared Diamond (1999), *Guns, Germs, and Steel: The Fates of Human Societies*, W. W. Norton & Co.
3. eMarketer Report (2013), Worldwide social network users: 2013 forecast and comparative estimates, http://www.emarketer.com/Article/Social-Networking-Reaches-Nearly-One-Four-Around-World/1009976 Accessed October 2013.
4. Adam Thierer (2013), 'Technopanics, threat inflation, and the danger of an information technology precautionary principle', *Minnesota Journal of Law, Science & Technology*, 14, 309–86.
5. Susan Greenfield (2009), *ID: The Quest for Identity in the 21st Century: The Quest for Meaning in the 21st Century*, Sceptre.
6. Phil Zimbardo (2012), *The Demise of Guys: Why Boys Are Struggling and What We Can Do About* It, TED publishing.
7. Independent Parliamentary Inquiry into Online Child Protection, April 2012.
8. http://www.theguardian.com/world/2010/mar/05/korean-girl-starved-online-game
9. Diana I. Tamir and Jason P. Mitchell (2012), 'Disclosing information about the self is intrinsically rewarding', *Proceedings of the National Academy of Sciences*, 109, 8038–804.
10. M. Naaman, J. Boase and C. H. Lai (2010), 'Is it really about me?: Message content in social awareness streams', Proceedings of the 2010 ACM Conference on Computer Supported Cooperative Work (Association for Computing Machinery), Savannah, GA, pp. 189–92.
11. Leif Denti et al. (2012), 'Sweden's Largest Facebook Study: GRI rapport 2012–3', https://gupea.ub.gu.se/bitstream/2077/28893/1/gupea_2077_28893_1.pdf
12. Amanda L. Forest & Joanne V. Wood (2012), 'When social networking is not working: individuals with low self-esteem recognize but do not reap the benefits of self-disclosure on Facebook', *Psychological Science*, 23, 295–302.

13. Robert B. Cialdini, Richard J. Borden, Avril Thorne, Marcus Randall Walker, Stephen Freeman and Lloyd Reynolds Sloan (1976), 'Basking in reflected glory: Three (football) field studies', *Journal of Personality and Social Psychology*, 34, 366–75.
14. Samsung poll and press release: http://www.samsung.com/uk/news/local-news/2013/samsung-nx300-wi-fi-every-day-over-1-million-photos-are-shot-and-shared-in-60-seconds
15. http://techcrunch.com/2012/08/22/how-big-is-facebooks-data-2–5-billion-pieces-of-content-and-500-terabytes-ingested-every-day/
16. M. D. Conover, J. Ratkiewicz, M. Francisco, B. Gonçalves, A. Flammini, and F. Menczer, 'Political polarization on Twitter', Proceedings of International Conference on Weblogs and Social Media 2011 (Unpublished, 2011), http://truthy.indiana.edu/site_media/pdfs/conover_icwsm2011_polarization.pdf
17. Sameer Hinduja and Justin W. Patchin (2010), 'Bullying, cyberbullying, and suicide', *Archives of Suicide Research*, 14, 206–21.
18. Vint Cerf speaking at the Consumer Electronic Show in January 2013: http://mashable.com/2013/01/09/would-you-wear-internet-connected-clothing/
19. Eli Pariser (2011), *The Filter Bubble: What the Internet is Hiding From You*, London: Penguin.

Index

Note: *italic* page references indicate illustrations

B

C

D

H

I

Q

R

T

Economics: The User's Guide

Ha-Joon Chang

What is economics?

What can – and can't – it explain about the world?

Why does it matter?

Ha-Joon Chang teaches economics at Cambridge University and writes a column for the *Guardian*. The *Observer* called his book *23 Things They Don't Tell You About Capitalism*, which was a no.1 best-seller, 'a witty and timely debunking of some of the biggest myths surrounding the global economy'. He won the Wassily Leontief Prize for advancing the frontiers of economic thought and is a vocal critic of the failures of our current economic system.

A PELICAN INTRODUCTION